AF695528

Innovation, Research and Development Management

Innovation, Research and Development Management

Patrick Gilbert
Natalia Bobadilla
Lise Gastaldi
Martine Le Boulaire
Olga Lelebina

iSTE

WILEY

First published 2018 in Great Britain and the United States by ISTE Ltd and John Wiley & Sons, Inc.

Apart from any fair dealing for the purposes of research or private study, or criticism or review, as permitted under the Copyright, Designs and Patents Act 1988, this publication may only be reproduced, stored or transmitted, in any form or by any means, with the prior permission in writing of the publishers, or in the case of reprographic reproduction in accordance with the terms and licenses issued by the CLA. Enquiries concerning reproduction outside these terms should be sent to the publishers at the undermentioned address:

ISTE Ltd
27-37 St George's Road
London SW19 4EU
UK

www.iste.co.uk

John Wiley & Sons, Inc.
111 River Street
Hoboken, NJ 07030
USA

www.wiley.com

© ISTE Ltd 2018
The rights of Patrick Gilbert, Natalia Bobadilla, Lise Gastaldi, Martine Le Boulaire and Olga Lelebina to be identified as the authors of this work have been asserted by them in accordance with the Copyright, Designs and Patents Act 1988.

Library of Congress Control Number: 2018945213

British Library Cataloguing-in-Publication Data
A CIP record for this book is available from the British Library
ISBN 978-1-78630-300-4

MIX
Paper from responsible sources
FSC® C013604
FSC www.fsc.org

Contents

Foreword

We live in a world that the Anglo-Saxons call the "VUCA world". In other words, our environment is characterized more and more by volatility, uncertainty, complexity and ambiguity. R&D (research and development) is no exception to this rule. While the strategic nature of innovation and its importance for the survival and success of businesses could hardly be questioned, the role of R&D departments and the modalities of their management remain an under-investigated topic. The following are the most common questions that arise: how does one organize R&D activities in the short, medium and long terms? How can one measure their effectiveness? What is the real added value and how could we measure the return on investment?

Globalization brings additional difficulties to R&D, as it questions not only the strategic positioning of this function inside organizations but also the very existence of the latter. Some ponder whether it would be more appropriate to refer to "C&D" (connect and develop) by exploring the resources outside of the company like start-ups or suppliers in order to avoid fixing long-term capital.

To clarify all these issues, in this book, we conduct a thorough analysis of the different types of R&D organizations, the recent advances in the field and the strategic choices made by different companies in transition.

R&D is a difficult area to define in a business because it is only partially subjected to the performance indicators commonly used by other functions like marketing or production. Even a basic notion like wastage is difficult to define in this function. To find new solutions, we have to explore new territories. Although this can lead to failures and waste, we must accept such wastage in order to innovate.

The R&D area poses many challenges to the human resources department in managing recruitment and competence development in a different way from other departments by implementing effective training plans to cover future needs.

The "H" in Human Resources is key. Companies are made by employees, by their skills and abilities and, above all, by their passion. It is easy to develop and implement the desired techniques and models, but ultimately the success of the R&D function depends on the level of passion that employees venture on a path, which will not provide a return in the short term as in other departments. It is thus of primary importance to build an R&D department comprised of employees who have a passion for innovation.

Roberto RENIERO
Agricultural Engineer
Doctor in Molecular Biotechnology
President of CIME

Acknowledgments

This collective work was supported by the Cercle de l'Innovation et du Management et de l'Expertise (CIME), a French association created in September 2016 with the aim of building and disseminating knowledge in the areas of R&D management, expert and expertise management and managerial innovation in France and worldwide.

CIME is a laboratory of experimentation and innovation supporting talent, experts and expertise management. It comprises:

– a group of company directors (R&D directors, Human Resources directors), practitioners and researchers interested in R&D;

– a managerial innovation group, consisting of HR directors, managers, researchers and practitioners interested in the field of organization and people management; and

– a resource center based on the results of action research, the sharing of experience and co-development.

CIME activities include sustaining and further developing the resource center with the content based on diverse studies, research materials and presentations (available at: www.cime-innovation-management-expertise.com), organizing meetings with peers and proposing opportunities for networking, granting access to first-hand expertise through the presentations of companies' representatives and academic researchers and capitalizing on the developed knowledge.

As of January 1, 2018, the CIME member companies are:

– Airbus Group;

– Boehringer Ingelheim;

– Crédit Agricole centre-est;

– Enedis;

– Groupe EOS;

– Groupe Saint-Gobain;

– Groupe SEB;

– Holding Textile Hermès;

– IRSN;

– Naval Group;

– Nestlé;

– Poclain Hydraulics;

– Schneider Electrics;

– Siniat;

– Ubisoft;

– Veolia; and

– Vinci Energies.

We are grateful to these companies for inspiring the writing of this book and contributing in many ways.

Beyond the current members of CIME, we would also like to thank all the other companies we have worked with in the last fifteen years, who have trusted us, accepted the sharing of experience, exchanged ideas and provided the necessary reflexivity for the progress of professional practices. We sincerely thank the directors, managers and R&D professionals of these companies for opening up for us their R&D centres, devoting their time and inspiring us to complete this book.

General Introduction

The objective of this book is to determine the ways of organizing, managing and leading the research and development (R&D) of a company. This General Introduction presents the book's aim, the issues associated with the R&D function and its management and the relevance of writing a book on the subject. First, we discuss the scope of this book in terms of types of organizations, business activities and issues, as identified by the authors. Second, we try to provide answers to key questions on R&D and innovation management by addressing the wider target audience of this book. Finally, the structure of this book is presented with a summary of the six chapters that cover many of the critical contemporary issues in the field of R&D and innovation management.

I.1. Managing R&D and innovation: current and critical issues

The current phase is marked by the increasingly strategic nature of innovation, particularly technological innovation, which has become central for developing not only competitive advantage of the company but also a crucial condition for the survival of businesses. Companies must continuously innovate, often and more quickly. Innovation results in new products that are completely different from the previous ones, and allows to offer a wide range of products to customers. One of the main reasons to innovate is to meet the expectations of customer renewal and to cope with the rapid outdating of products due to technological progress and competitors' strategies. It is important to both innovate more quickly and consume less resources so as to ensure certain profitability of innovations despite the simultaneous rise of design prices and the growing risk of technological obsolescence. Chrono-competition and the *time to market* have become key dimensions today in business competitiveness, which strongly influence innovation process management.

In industries based on science and technology, where product and process innovations are technological in nature, the R&D function is a major player in the innovation process. In the current context marked by competition through intensive innovation [LEM 06], R&D is thus placed very high on the agenda of science and technology firms.

Indeed, the purpose of R&D is to develop new scientific and technical knowledge and to use it in the design of technological innovations, which may consist of new processes, products or services that bring added value compared to existing solutions and offers. R&D enables the opening of research avenues and for one to pursue them internally for harnessing the available product knowledge; it comes up with new technical solutions; proposes company product and process improvements and technically assures and coordinates their development. This function can be distributed throughout the company or be incorporated in a specific area. In the first case, it is led by different experts, engineers and scientists who work in the production services or in technical support. In the second case, the company creates a specialized unit dedicated entirely to R&D activities, which consists of diverse specialist teams. In any case, R&D plays a major role in not only driving but also initiating technological innovation, and, as such, it holds a privileged position in strategy definition, alongside marketing, production, procurement and so on. In 2015, companies in France invested 31.8 billion euros and employed 251,444 people in the equivalent full-time in R&D.

While innovation has become a key competitive advantage in a number of business activity sectors, the role of R&D in innovation processes varies considerably depending on the sector: people services, the world of fashion and clothing, the motor industry, the aeronautical industry and so on. For high-technology industries, R&D is essential in the innovation process and it can be considered as a standard support function. Similarly, in the domains of health and biotechnology, the spatial domain, electronics, the semi-conductor industry, informatics or communication technologies, R&D is the core of a company and is directly linked to its ability to renew business proposals and conduct its strategic activities. In industries of lower technological intensity, R&D is less directly related to the actual capacity to produce a business proposal, but it contributes to competitiveness by bringing differentiation elements/unique selling points to the products in order to compete nationally and internationally[1].

However, although technological innovation and R&D functions are more strategic than ever, R&D is losing stability at present due to difficulties associated with the acceleration of innovation processes in a changing and openly competitive world. R&D finds itself in a difficult situation and is more and more questioned by managers and company directors. How does one define and measure the performance and therefore profitability (particularly for the shareholders) of investments in R&D? Could it be possible to increase efficiency and effectiveness by streamlining working methods and, consequently, enable a reduction in R&D expenditure while increasing their contribution to innovation processes? Is it really necessary to keep the R&D centers on national territory when the markets and strategies are becoming international and when localizing R&D in the "emerging" countries combines the advantages of proximity to the growth markets and low wage costs of scientists and well-trained R&D engineers? Do we really need to continue the R&D activities ourselves in-house, despite recent trends toward open innovation and offshoring, which aims to capitalize on resources in terms of ideas, knowledge and skills from a myriad of very different players (start-ups, customers, experts, consultants, etc.) everywhere in the world? R&D is trapped in these critical and global changes in the governance models and is in permanent search for new organizational arrangements allowing us to tackle the diverse managerial issues and handle the ever-changing professional identities.

However, management, both as a discipline and as a set of practices, is almost helpless when it comes to managing activities characterized by uncertainty, the creativity issues, specifics of the professionals that have more loyalty and attachment toward their profession than their organization and so on. The management techniques developed to manage and control the more repetitive tasks of production and operation often appear inadequate in the context of R&D. It is thus necessary to have specific knowledge on the

1 These differences are particularly noted at a level of R&D intensity that varies from less than 1% of the turnover spent on R&D in sectors such as transport, construction, textiles, paper or oil, up to 10% in the fields of e-maps, computers, medical equipment and measuring and navigation devices. Sectors such as chemistry, metallurgy, motor and naval construction and telecommunications have R&D intensity in the range of 1–5%, whereas it is between 5–10% for pharmaceutics (source: Ministry of Research and Higher Education, 2014).

management methods within these organizations, the thorough apprehension of their activities and the understanding of the specificity of R&D professionals.

Although all of the political, economic and academic players agree to recognize the fundamental role of the R&D function in the innovation process and in the growth of the company, more knowledge needs to be produced to better understand the complex processes and organizational issues that lie behind R&D activities.

This ambition was a starting point that motivated the creation of this book. All the authors are management scholars and practitioners who have similar interests for the world of R&D and technological innovation. For many years, they have been sharing the results of their research and management experiences both between themselves and with the senior management of companies (technical and R&D managers, experts, scientific directors, HR) guiding them in their strategic choices and managerial queries. The authors intended to spread this accumulated knowledge to a wider audience and to raise some questions that could be considered "off the radar".

I.2. A book on R&D management of large industrial enterprises

This book seeks to reveal the critical issues and practices of modern R&D management in companies and addresses its current numerous challenges and confusions. It is a broad subject and we had to make some choices.

First, we must ensure that the central subject of this book is R&D and its management methods, analyzed in combination with the current issues in innovation. It is not a book on innovation management per se. The innovations addressed in the following chapters are mostly technological, which is only one type of innovation among many (marketing, commercial, organizational, managerial, etc.), and even with regard to technological innovations, we do not cover all the questions related to their management. Thus, the financing of innovation, market analysis, the marketing of innovative products and spin-offs do not fall within the scope of this book, which focuses on R&D management as one of the major contributors to the process of innovation.

In terms of the companies that we have taken into consideration, the focus has been placed principally on large industrial companies. There are, however, three limitations. The first is the exclusion of academic research activities carried out in institutions of higher education and research centers (universities, Grandes Écoles, engineering colleges, research laboratories, research foundations), whether public or private. We focus on the R&D of companies that design, produce and market goods and services destined for customers in competitive markets, both oligopolistic and monopolistic. This being said, we consider the connection between the R&D of these companies and the world of academic research, often referred to as science–industry links.

The second limitation of our analysis concerns the size of the companies that we have included in our study. Indeed, we focus on large and often multinational corporations as they conduct the majority of R&D activities, as evident from statistics. Data from the French Ministry of Higher Education and Research shows how in 2014 companies with less than 500 employees (therefore including companies of a significant size) represent 94% of French companies and only 41% of the R&D workforce and 33% of R&D expenditure. Only 6% of the total number of companies have more than 500 employees, but they provide 67% of the total expenditure on R&D. Within these, companies with more than 5,000 employees are very few (0% in the statistics due to rounding!), yet they account for 25% of the workforce and constitute 31% of R&D expenditure.

The choice to focus on large businesses rather than on more modest-sized enterprises is also because R&D units are created first in large organizations. These companies can therefore invest large amounts in activities with uncertain and often delayed profitability, which is more difficult for smaller businesses that have fewer resources available (especially before they could benefit from supportive instruments such as start-up funds, business angels and venture capital). These are also the companies that have already experienced difficulties with R&D management and have come up with possible solutions. Therefore, these companies build on practices that are long-standing, sophisticated, detailed and formalized.

As a result, they have also been analyzed more by the researchers and consultants and are now part of an extensive and stable knowledge base.

These data do not mean that there is no R&D in small and medium-sized enterprises (SMEs). Indeed, even very small enterprises can conduct R&D activities and experience difficulty in both the daily management of such activities and the implementation of creative policies. The impressive development of start-ups in recent years has really shifted the focus toward small, technology-intensive companies basing their innovation on R&D capacities. In this book, although we choose not to focus on the latter, we will still deal with start-ups and their specific characteristics. It is also worth noting that large and small companies have a reciprocal interest and mutual inspiration practices, even though the latter will inevitably be somewhat different due to the disparity in their size, structure, economic and financial model, the available resources and so on.

The third and final limitation of this study involves the choice we made to concentrate on large companies that operate in the industrial sector. Indeed, even if a vibrant R&D can exist in other sectors (especially in business services), the major design activities are concentrated in the industrial sector (or industrial sectors considering their great diversity)[2]. We must note, however, how rankings and boundaries become blurred and penetrable. The field of IT and communications is emblematic of this trend as it crosses the secondary and tertiary sectors, designing, using and building on infrastructure, products and services. We can observe this trend, to a lesser degree, in many other sectors. The motor industry, for example, demonstrates the fact that at present, in addition to manufactured and marketed products, a variety of services are offered to clients (communication technologies, banking, insurance, after-sales service, etc.). If this book is biased by focusing on industry (motor, chemical, pharmaceutical, electronics, IT and communications, energy, etc.), then it is worth mentioning that the boundaries of the latter are already and will be even more transparent over time.

Within this defined framework, we have tried, however, to deal with the subject considering the entire array of activities, which are covered by the acronym and which vary dramatically based on whether it is research or development. We also tried to include the analysis of diverse activity sectors

2 According to the nomenclature used by the French Ministry of Higher Education and Research, in 2015, 73.5% of internal spending on R&D in French companies was by manufacturing industries, 4.6% by the primary sector, energy and construction and 21.9% by the services industry (a large part of which is the work of companies offering scientific and technical services).

and technological fields. The boundaries between what could be considered as R&D (and what could not) are blurred and depend heavily on the choices that companies make in terms of their internal organization. This is reflected in the work of Frascati, prepared by the OECD, which sets out the criteria to decide what statisticians should consider being within the perimeter of R&D. It stipulates that R&D (experimental research and development) encompasses work that is done systematically in an effort to increase the knowledge base, including knowledge of people, knowledge of culture and knowledge of society, as well as the potential interest of this knowledge base for new uses. This excludes from the scope of R&D other activities such as the production and marketing of innovations, as well as scientific and technical activities that differ from R&D despite their close connection, such as education, training or patent-related legal work.

Thus, in our analysis, we provide a vision of R&D in its entirety and its diversity, by trying to avoid the pitfall that could be observed in a number of research works and which treat R&D as a homogeneous entity. Therefore, we consider that business logic, professional identities, the purpose and nature of business activities as well as the types of management could vary significantly between research and development. We believe that both have different meanings and practices depending on the scientific domain (exact and natural sciences, technological and engineering sciences, medical sciences, agricultural sciences, etc.).

I.3. A descriptive and analytical book aimed at a wide audience

As the objective of this study is challenged by numerous issues and boundaries, it is helpful to define the context in which we will situate our analysis. Without claiming to cover the full range of practices, we wanted to capture the reality and the diversity of the latter. This diversity relates both, as was noted earlier, to the variety of activities regrouped under the terms R&D and innovation, and to the variety of business sectors. It relates also to the specific characteristics in each type of business, which leads to choices in terms of governance, structure, individual and team management and so on, which could differ even within the same business sector. We will illustrate our theoretical and practical developments through testimonies and case studies. The content is based on our knowledge, available public data and collaborative research carried out in various R&D organizations.

These illustrative examples will sometimes be accompanied by critical comments: all practices, even those that are commonly used and very far-reaching, may not necessarily need to be applied and promoted. On the one hand, certain practices have shown their limitations and we will discuss these, requiring the invention of new R&D management methods. On the other hand, we are convinced of the need to contextualize company responses to the difficulties and challenges of R&D and innovation management. Not only do the problems but also the contexts in which these problems occur differ from one company to another – this requires a varied set of answers adapted to each particular situation. Therefore, the objective of this book is not to propose predefined solutions, but rather analytical frameworks and examples, which could serve as inspiration and should not be simply copy-pasted.

This book – *Innovation, Research and Development Management* – will be useful for all those willing to deepen their knowledge of the subject, whether in the academic or professional arena.

In the academic sphere, this book is aimed at students as well as lecturers and researchers. We hope that Master's and Doctorate students in universities as well as in business and engineering schools will find the content interesting and useful. This book could also be of particular interest to those who consider pursuing a career in R&D: R&D engineers, managers, project managers, HR professionals working within R&D units and so on. Finally, this book is also aimed at lecturers and researchers who wish to have an overview of the field through specialized articles on the major key topics of R&D management and technological innovation. Taking into account the issues developed in this book, it may also be of interest to researchers working on similar subjects, such as the management of creative activities and project management.

Furthermore, this book could be useful to professionals, whether they are R&D researchers and engineers, R&D managers (whether in operations or heads of specific R&D programs) or managers of various support functions working with R&D (HR, accounting and finance, etc.). It provides the essentials on the specific characteristics of R&D, the challenges and difficulties of its management, examples of practices and critical issues to be taken into consideration.

At the intersection of public and academic audiences, this book can serve as training material destined for employees and R&D managers, as only few studies have been published so far in this area.

Also, because modern companies are operating in a knowledge-based economy, this book could be of potential interest to all those concerned with the management methods of knowledge-intensive organizations that employ knowledge workers.

Finally, we hope that it will attract the attention of public policy makers who, being anxious to promote and accelerate innovation as a major condition of national competitiveness (through instruments such as competitive clusters and future investments), seek to better understand the innovation process and the R&D activities.

I.4. Structure of the book

This book contains six chapters offering a wide overview of the key aspects of R&D management, even if we are not claiming to be exhaustive. Indeed, we have not devoted any chapter to R&D information systems, the management of industrial property or the subject of the CSR (corporate social responsibility) in R&D.

Chapter 1 concerns the environment and strategy and lays the foundation. It focuses especially on the emerging issues in R&D that are highly influenced by globalization. We learn that ongoing transformations are instigated by a combination of factors: increased technological changes, pressures exerted by the market and the evolution of R&D strategies. The result of these influences is a strong tension between the need for innovation and the need to control costs in the context of R&D globalization. This chapter also shows how influential geostrategic logics in R&D (traditionally a Western domination) are currently on the way to being reversed. We will go into this question in greater depth by studying the problems that result from R&D internationalization in major multinational companies.

Chapter 2 focuses on the specific characteristics and the transformations in R&D work. Although the latter may be distinguishable from the activities of the other major functions of the company, it is influenced by profound changes that have emerged since the 1990s and that are far from being terminated. This chapter identifies both the factors and consequences of these transformations. Therefore, beyond the developments in R&D work that came along with scientific progress and technical expertise, competition through innovation (which has currently become the golden rule in many technology sectors) influences R&D activities. More precisely, it is a combination of events that modify and destabilize research and development work, including the rise of project management, pressure to reduce the time-to-market ratios, rationalization of the innovation process, user-centered design, open innovation, internationalization of R&D activities and digital revolution. This, in turn, has an important impact on the positioning of R&D in the organization, its role in strategic decision-making, the objectives and expectations from this function, the time frame of the work, the degree of autonomy attributed to the R&D researchers and engineers, the working conditions, the competencies required in R&D and so on. This chapter raises a number of questions about the future of R&D's work and, in addition, the future of the R&D function in modern organizations.

Chapter 3 focuses on rationalization strategies in R&D and their effects on creativity, while highlighting the notion of *slack*. It outlines the main factors involved in innovation and the creative process at the heart of R&D organizations. Then, the chapter describes the different types of slack reduction strategies and how rationalization of R&D resources (human, financial, time, space) affects the creativity of professionals and teams. With drawing on the case of R&D organizations being the subject of important rationalizations over time, this chapter highlights the elements involved (procedural, relational, emotional, cognitive, etc.) and how these elements influence the creative process (generation of ideas, sharing ideas, intrinsic motivation) of individuals and teams.

Chapter 4 deals with current practices and challenges in the area of human resources management in R&D. In the new millennium, management of R&D professionals has been the subject of ever-increasing interest, both empirically – with R&D organizations being engaged in important projects in the field of human resources management (HRM) – and theoretically. This chapter focuses on the analysis of HRM practices in R&D and their evolution in relation to major changes in work organization. The objective is thus twofold. First, we highlight these practices, their emergence and their concrete modalities. Second, our objective is to question their functionality as follows. What are the advantages and limitations of these new HRM and R&D practices? Do they meet the constraints and the challenges that R&D organizations are facing today? What are the points of vigilance where it is important to engage R&D professionals whether they are technical managers or HR managers?

Chapter 5 focuses on collective expertise, in a way that can be seen as contrasting to Chapter 4. In fact, although the majority of studies consider knowledge as an individual attribute, here, we explore the collective aspects of knowledge and its different forms of structuring. This chapter first presents transversal expertise communities (troubleshooting teams, research projects, panels of experts) and discusses the prospects and challenges for a collective endeavor by R&D professionals within these communities. The willingness to pass across the boundaries (of expertise domains, functions, departments, etc.) is currently a major preoccupation of large corporations. However, we emphasize that too much of such a multidisciplinary approach may bring the risk to the preservation and development of specialized knowledge and result in an "organizational forgetting" of the crucial information. To address this risk, we draw attention to the need to analyze and manage the dynamics of knowledge and the interplay of roles of R&D professionals working within the same expertise domain. Finally, we highlight the importance for companies to distinguish as well as structure these two forms of organizational expertise (transversal and domain-specific).

The last chapter, Chapter 6, describes and questions both the notion of R&D performance and its governance. It is a tricky subject for many reasons. Some are linked to the "status" of R&D and are similar to various support functions: these include difficulties regarding performance measurement and the adjustment between the resources used and the economic performance. Other factors are intrinsic to the field of R&D and innovation: this applies mostly to the particular nature of innovation processes and to the uncertainty in the research activity results. However, the financial crisis has deepened the focus on the control of R&D performance and, consequently, specialists have attempted to find both quantitative and qualitative indicators of such performance. This chapter presents and analyzes the different existing management performance practices, while proposing guidelines to choose the correct performance indicators in the area of R&D.

The book concludes with an abundant bibliography that guides interested readers through a number of relevant and recent references in the field of R&D management and technological innovation. Finally, this is followed by an index identifying the key concepts discussed in the various chapters, as well as the names of the companies that serve as examples throughout.

1

R&D and New Competitive Challenges: Between Intensive Innovation Strategy and Internationalization

R&D activity has always occupied a central place in the strategy of large enterprises and in the competitive performance of states. Recently, with a significant change in the competitive environment of companies (such as, for example, competition with new emerging countries, intensive innovation policies, changes in customers' expectations and transformation of business models) innovation and R&D have become essential for both company and national competitiveness.

In this chapter, we first present the relationship between R&D and company strategy and the different generations of R&D and their characteristics, which are currently possible to identify over a long period of time. We then identify and describe the main factors that influence R&D strategy in companies, emphasizing their impacts on the construction of new R&D models. Finally, it provides a practical illustration of the latest developments in the R&D sphere through the presentation of policies that Western companies have implemented in their Chinese subsidiaries.

1.1. Strategy and R&D

The place of R&D in the company is closely linked to technological innovation, particularly in its strategy. It is important to note that innovation is not simply an invention, but it is the result of a design process that a market has endorsed. Ultimately, it is about successfully bringing to market profitable requests (products or services), for which the company undertakes a reshuffle of knowledge and existing skills in order to offer new products or services. The place of R&D also depends on the sector, competitive strategies and the nature of the company's markets.

1.1.1. *R&D's place in business strategies*

R&D plays a special role both in companies' strategies and in the international arena. Research and development projects not only provide a competitive edge for businesses, but are also vital for the economic competitiveness of countries in their struggles for power and international influence.

1.1.1.1. *R&D responsible for its very unique place in company strategy*

The place of R&D in company strategy depends, first, on the business sector, the nature of the markets and the company's competitive positioning and, second, on its degree of globalization. Therefore, the high-technology companies belonging to the "innovation-driven" sectors (as well as pharmacy, aeronautics, energy, electronic, etc.) [GIR 96] make huge investments in R&D, whereas the businesses in mass distribution invest little or no funds in R&D. The nature of the markets also affects the position of R&D in company strategy. For consumer products manufacturers, the R&D strategy will consist of manufacturing better and cheaper than the competitors, by offering either innovative products that better meet consumer expectations or products with new and highly attractive features (e.g. the new digital applications via smartphones or social networks) or by adding a service to the product (such as integrated remote maintenance with the purchase of a printer). For these companies, R&D activity has long been closely linked to decision-making. When dealing with the B2B sector, that is, those geared toward the sale of products to industry, a large part of R&D is performed close to industrial entities, where the industry is located [LEN 02], in order to respond to the needs of cost reduction and to support innovation efforts. This is done by providing "solutions", on the one hand, and by creating innovative concepts capable of introducing breakthroughs in industrial performance on the other. This is the case, for example, with R&D centers of the cement industry being located near production plants.

The strategy of localizing a company's R&D activity close to its customers is illustrated by the example of Usinor, now known as ArcelorMittal.

Usinor (ArcelorMittal, from 2006) is a company producing raw material (steel), involved in the upstream of industrial clusters, particularly the automobile industry. The innovation strategy, which operates in the automotive market, is also reflected within this upstream company, which is forced to find innovative technical solutions, which are at odds with its dominant position on the market.

Here, the proximity, not only geographical but also organizational and institutional, between the R&D of the upstream company and car manufacturers is a prerequisite to its success. Organizational proximity involves putting in place coordination and cooperation arrangements between R&D players of the iron and steel group and those of the car manufacturers such as shared research projects allowing the exchange of information and knowledge. In contrast, institutional proximity involves creation of common areas of co-exploration for interactive learning required for innovation.

Besides the traditional strategic aspects like the domination by the costs and benefits of global "solutions" for the client, this company's R&D strategy has always been about developing and proposing innovative product proposal (IPP) projects in the field of hydroforming, in which:

– the result of the design project is not linked to a specific object designed directly for the customer;

– the project is likely to shake up the mainstream R&D representation of the company and its customer;

– uncertainty about the customer accepting the innovation is the norm.

These innovative product proposal projects lead to a situation of new cooperation with customers referred to as "co-exploration", in which the two partners explore together, without certainty as to outcome, the concepts likely or unlikely to lead to the creation of added value.

Box 1.1. *Design differentiation approach in the R&D of Usinor (source: Lenfle and Midler [LEN 12])*

The place of R&D in company strategy also depends on the globalization of the business activity. In order to compete, some companies had no other choice but to globalize. The analysis of globalization provided by the economist Pierre-Noel Giraud [GIR 12] serves as essential reading that helps us understand its effects on the breakdown of the value chain and, in particular, on R&D's place in globalized enterprises. The Giraud model is based on the division of the world economy into territories separated by borders. It is based on a distinction between "nomadic firms" and "sedentary firms" and on a differentiation between the jobs of a territory that produce goods and services subject to international competition and those producing goods and services protected from this international competition.

Within each territory, protected by borders, people, capital and goods circulate freely, but not always between territories. This concerns human resources in particular. The nomad firms produce and circulate economic objects between territories. They make territories compete in their comparative advantages to make profit, and their decisions on FDI (foreign direct investment) destabilize or stimulate the territory's economic activities. This is particularly the case with R&D investments. The sedentary enterprises only produce or trade at the national level. Employees competing for the nomadic firms will also be described as nomads. They only compete on a territory if they are competitive, otherwise they are transferred elsewhere. Sedentary or protected employees only compete with one another for the production of goods and services for local use. Thus, the global firm "locates sedentary activities where they must be", and the nomadic activities "in the territories that offer the best conditions": low labor costs with high labor intensity and availability of highly qualified scientific staff for research laboratories.

By repeating Giraud's paradigm, it is clear that R&D's position is of crucial importance for companies from overseas or nomads, that is, those exposed to global competition, that are on the front line of innovation and product renewal; it is less crucial for "domestic" companies or those on the protected market like mass distribution.

1.1.2. *Different generations of R&D*

The governance of R&D is strongly impacted by the level of importance that the company attributes to innovation. Other factors are included in the connection between the development and creation of scientific and technical knowledge and value proposition for customers.

Management literature has highlighted five generations of R&D management [ROT 94], which are described as follows.

The *technology push* model was the first dominant generation during most of the 20th Century; it is very visible in the policies of R&D enterprises as well as in those of states through large research organizations, such as the CNRS and the CEA in France or NASA in the United States. According to this model, innovation is primarily dependent on the industrial and scientific policies of companies or states according to a top-down view (Schumpeter). Scientific and technological progress guides innovation. The main participants are the researchers and engineers in R&D departments of large companies or scientific organizations. This does not mean it is a closed model. While public–private exchanges are not a particularly recent tendency, openness and networking continue to play major roles in the world of science and technology.

This approach, also referred to as the "producer's model" [VON 01], has been of major importance in the development of areas such as telecommunications, aerospace, nuclear science and robotics through government policies. In enterprises, new products and services are designed and developed by the R&D team and then promoted and sold through the marketing department to be ultimately purchased by the consumer. Innovations can be incremental (i.e. small scale that build on what already exists) or radical (large scale or breakthrough, i.e. they offer solutions that did not previously exist). This model has also been described as an "Emitter innovation model" model [BAD 13] in reference to the fact that, based on a technological determinism, innovation is designed and released to the company or the markets in a linear and mechanical way with a top-down view.

The *market pull* generation emerged toward the end of the 1960s. Building on the consideration that the consumer is essential in defining needs, it combines the logic of supply in R&D based on technology and the logic of demand resulting from the increase in consumer expectations. We start from market needs and expectations to adapt the technology according to the latter. The marketing department, which studies subtle market needs, is central to the arrangement. The R&D function is the in-house IT service provider for the marketing department and senior management, which provides technical solutions to operational needs.

The years 1970–1980 saw the advent of a third R&D approach which, despite the saturation of markets, began to introduce rationalization efforts through cost reduction and control and through measuring all R&D investment against companies' business strategies and financial expectations for return on investment.

The end of the 1980s was characterized by an R&D approach in which it then became important to merge strategies that intersect the rationale of supply determined by technology and those of consumer-driven demand. This is an integrated approach and crosses several of the enterprise's functions like marketing, distribution and customer service. This characterized the fourth generation of R&D management.

Finally, the 1990s saw the arrival of a radically new design. A company's knowledge capital was to give it a sustainable competitive advantage in its market in terms of its capacity to generate product or service innovation. The enterprise was witnessing the rapid disintegration and outdating of knowledge of its professions and technologies. In the era of "time compression" [ROS 10], short time-to-market products needed the innovation process to speed up. To address very complex innovation problems, the need for R&D was advancing. Enterprises conducting R&D activities in-house were then obliged to rely on an external R&D ecosystem. Consequently, they established more open innovation design practices. R&D then introduced a multitude of internal and external practitioners to the business or organization, maintaining links and relations that might contribute to innovation. It is the generation of *open innovation* [CHE 03].

In a background of intense and accelerated competition, Western companies have no other choice but to get a maximum return on R&D investment by resorting to open innovation, allowing them to take advantage of the network of skills of many partners to speed up new product development and add value. This partnership can sometimes be seen elsewhere with certain competitors as can be seen in the pharmaceutical sector, where R&D efforts to ensure the development of a really innovative and profitable molecule (blockbuster) require huge financial investments (on average, several hundreds of millions of dollars). Enterprises such as

Sanofi have developed, for these reasons, a strategy of open innovation for the whole of the group in a B2B approach.

In the pharmaceutical sector, two main participants coexist: major groups who were historically based on knowledge and know-how coming from chemistry and the smaller firms more inclined toward biotechnology. Public laboratories also play a key role in R&D and innovation in this sector. The firms in the sector are directly involved in scientific advances in biomedical science (science-based sector).

The Sanofi group, the third largest pharmaceutical group in the world in 2016, started to note in 2008, as in the whole of the sector, a decline in the productivity of R&D and adopted a model of open innovation to broaden its resource base in knowledge and know-how and increase its capacity for innovation. It is necessary for the group, at the same time, to reduce its dependence on blockbusters (molecule likely to give a decisive advantage), in order to master the threat of generic drugs, which are increasingly attractive to the national health authorities, and optimize its R&D.

The strategy was to enroll in a three-party collaboration, in which academic laboratories lead public basic research; biotechnology companies lead the upstream phases of research; and the major groups are responsible for the development and commercialization phases. This strategy became a reality for the group, between 2008 and 2014, through 26 generic biotechnology acquisitions (the most important of which was the biotechnology Genzyme) and vaccines to incorporate innovation in the pre-clinical or clinical phase in R&D. It is complemented by the establishment, around the same time, of 104 partnerships and alliances to enrich the knowledge base of the group on topics that were inadequately controlled.

In this model, the policy of open innovation, which focuses on both targeted partnerships and a policy of company acquisitions that develop specific molecules before going to market, enabled a form of outsourcing of exploratory R&D (the R of R&D) to help them acquire a true "pipeline" of molecules that have already been tested and are likely to convert to commercially profitable molecules.

Box 1.2. *Sanofi's open innovation strategy (source: Labrouche and Kechidi [LAB 16])*

This is also the approach developed by a telecommunications operator like Orange, which aims to take a leading position worldwide.

The Orange Group is a global operator of telecommunications with a large R&D function (5,000 people across the world), particularly in Europe, where some operators have decided to withdraw (British Telecom, for example).

It is organized around the Technocentre in France, "a manufacturing and design facility", Orange Labs that have overall responsibility for technical products and services, a development center that identifies innovative trends in Silicon Valley and an Orange Valley dedicated to rapid innovation.

Orange has chosen to focus on the following three levers of innovation:

– Immersion in an innovative ecosystem: SSII, digital agencies and start-ups allow it to accelerate its innovation process by providing its customers with unique solutions and user-friendly access. Each of the group's research and innovation centers is integrated in a geographic ecosystem as close as possible to the local markets and is involved in public research programs, such as the investment programs for the future, with public–private partnerships or competitive clusters. Orange Labs carry out monitoring activities in Asia and the United States. Orange Institute facilitates the understanding of digital transformations for researchers and decision makers to external companies. Orange accompanies the external companies in a co-design approach, in particular, on the Internet of Things (IoT).

– Opening up of Orange assets to the ecosystem of developers through the application programming interface that makes 15 application programming interfaces (APIS) available on strategic projects for the group.

– Active support for start-up development with Orange partners (Orange Fab, start-up accelerator, Orange digital ventures and Orange Social Entrepreneur Prize). It supports an innovative workplace for the company employees and associates (NUMA, digital canteens), involvement in the French Tech initiative for valuing the digital ecosystem and promotion of intrapreneurship.

This wide-reaching program aims not only to associate the innovation capacity of young enterprises and innovators of all kinds to Orange expertise to accelerate the development of new solutions for customers, but also to open the doors of the enterprise to advance the images its own employees have on innovation in particular.

Box 1.3. *The open innovation of Orange (source: Orange [ORA 16])*

In this last generation of R&D as is illustrated by our two examples, sharing of information and knowledge between networks, clusters, start-ups, centers of public or private research, customers and suppliers, organized in ecosystems, appeared to be powerful levers for innovative design. In order to continue, it is the company that must advance its R&D by being able to use external ideas as well as internal ideas and take both internal and external paths to market [CHE 06].

1.2. Environmental factors influencing business strategies in R&D and their consequences

Having distinguished the different strategies witnessed in R&D matters and the latest trends, we must highlight a few significant factors that seem to us to significantly affect the current context in which R&D policies evolve in global enterprises.

We have made the choice to emphasize three of them that, according to us, exert a major influence on R&D strategy: the central role of innovation, the increasing importance of the consumer and the effects of market globalization.

1.2.1. *The major role of innovation in competition strategies*

Until the 1990s, R&D, examined from the geostrategic point of view, enabled industrialized countries to maintain a "competitive advantage" thanks to the design and production of high-added-value/high-end goods. These goods required a hi-tech innovation. They then had to be transferred to emerging economies, hoping to aid development and, in this way, enable the development of a global "virtuous spiral"[RIE 08].

The intensification of international competition and the growing roles of new economies and emerging markets led Western companies to reconsider their competitive conditions.

The arrival on the markets of new economic players from emerging countries (BRICS[1]) since the mid-1990s, coupled with the new financial

1 BRICS is the acronym for the five major emerging economies: Brazil, Russia, India, China and South Africa.

efficiency requirements of capitalism, has toughened economic competition while changing the place of innovation in firms.

First, innovation has become not only a means for achieving growth, reserved for certain sectors, but a crucial condition for survival, in particular through the policies of intensive innovation. In order to be competitive, it is necessary to differentiate oneself through gaining market knowledge and being attentive to the markets. It is also necessary to translate this into permanent innovation and to position oneself in the new sectors of business activity with a high degree of scientific and technical knowledge.

Second, for an increasing number of businesses, it has become critical to know how to identify and then capture the value created by harnessing knowledge and creating new knowledge in order to further their strategy and develop a competitive advantage. This requirement takes place in a context where market transformations and customer expectations are increasing and require more skill.

The example of the innovation approach of the Saint-Gobain glazing division shows how a company that has traditionally focused on a high level of technological innovation in its products is forced to advance toward the creation of value through solutions that provide benefits and results for clients.

The glazing division of the Saint-Gobain group experienced for many years an initial state of "dispersed and isolated research" in its automotive, windscreen and glazing products. This corresponded to a long period during which its product was relatively frozen. What prompted a reconsideration was the request from the car manufacturer Renault for the supply of a heat-insulating windscreen glazing, which would prevent excessive heat from the sun, so as to save on air-conditioning.

To respond to this, Saint-Gobain established a project team with a dedicated leader and contributors in the identified fields to manage the technical innovations to meet this request. However, other requests arose, such as heated glazing and the addition of an aerial, which made coordinated management of different projects and, especially, the control of their costs difficult.

R&D was then structured in "lines" of necessary skill sets to reuse the skills that have already been acquired in future projects. The lines enable the design of "semi-finished products", generic designs with a certain amount of freedom so they can be adapted to another product development. Different types of projects are incorporated – development, exploration, new technology – the design of which fall under different research rationales. Managing a line is no longer about having "off-the-shelf" technology, but about the development of organizing skills that will nurture new projects in a context of "repeated innovation".

Progressing to "intensive innovation", the company developed a new "glazing theory", known as a membrane, which both insulates and connects, with the hope of pursuing new features. It is not about innovation for a single client anymore, but about pursuing new areas of knowledge that could fuel new explorations.

Box 1.4. *Innovation strategy and management of R&D at Saint-Gobain Sekurit (source: Le Masson and Weil [LEM 02])*

The example of Saint-Gobain shows that the traditional criteria for innovation performance (essentially parametric and based on size, cost, efficiency and product safety, according to defined need) have also flipped; "the character of goods and services has therefore become unpredictable" [LEM 06]: a windscreen is also a smart communication media, a mobile phone is at the same time a camera, a watch is also a computer and an oven is at the same time a library of cooking recipes. Even if this new approach to innovation requires research in cost control (Chapter 6), the need for intensive innovation does not negate the tangible elements to justify R&D performance.

1.2.2. *The emergence of the consumer in R&D*

In the background of the changes just described, the problem is no longer only "to introduce a new product" to the market, a model that could really have meaning in an economy of technology-driven supply or market-driven demand, but to provide a product that meets a range of different need and practices and the search for other types of networks for potential customers or even knowing the importance of considering the capacity for innovation of the customer.

The subtle identification of customers' needs was, up to this point, the prerogative of marketing services who interpreted these analyses as operational specifications for R&D. However, understanding customer expectations swamped with numerous requests and occupied by new concerns can no longer be based simply on "questioning" them on their needs, because they can only develop a need based on the solutions that they already know or imagine.

The challenge is therefore to understand them better and to anticipate what they go through, as well as their difficulties and culture. The approach that prevails now is that of "end-users", where it is not so much about asking the customer the question "what innovation could my enterprise offer that would help you overcome the problems you are facing?". Rather it is to observe carefully what customers are often not able to express themselves because of one-sided understanding, habits, lack of knowledge on current technological possibilities and so on. R&D then uses methods that further resemble those that a sociologist, an ethnologist or an anthropologist would use rather than the traditional distribution of a satisfaction questionnaire.

In order to identify the innovation opportunities, the R&D function needs to adjust its possibilities and resources to match the observed client needs. "Customer research" should be about identifying clients who are specifically experiencing the problems that can be solved by the proposed innovation and for which this innovation "makes sense".

The Nestlé group, which has, for the past few years, aimed to develop a *consumer-centric* R&D approach illustrates this approach.

The Nestlé group, the largest group in the nutrition, health and well-being sector and founded more than 150 years ago, has always placed innovation at the heart of its business model. Henri Nestlé created the company starting with the invention and marketing of the first formula milk, aimed at newborn babies whose mothers had difficulty breast-feeding.

Like different generations of the R&D of global enterprises, the group successively directed its innovation efforts first toward the supply of food products and then toward nutrition and pleasure nutrition. It defines its innovation strategy currently as being geared toward nutrition, health and well-being, that is, enabling consumers to access healthy products in order to maintain good health and prevent diseases.

At Nestlé, the need for interaction with consumers transcends its only marketing function of study of consumption and customer service, to which it has traditionally devoted its efforts. The following points show how customer focus is also a necessity for the R&D department:

– by putting the research and development teams along with those from marketing in direct contact with the consumers (observation in purchasing and sampling, focus groups, immediate collection of feedback, co-creation within pop-up stores ,etc.);

– by promoting quick prototyping of a new product, based on the formation of joint teams blending the innovation approach, R&D function, marketing and the strategic affairs units.

The challenge is to lead all major players of these functions to set about their work in an unprecedented way, "in the consumer's position" or by putting oneself in the consumer's shoes.

This approach, which is unusual in R&D, borrows the methods and tools of design thinking (ideation, test and learn, etc.) and "learning by doing" most commonly used in marketing. It is based on the need to move the R&D professional's frame of thinking while allowing them to be open to new consumer approaches.

Box 1.5. *The Nestlé Group and the "Consumer-centric" approach to R&D*

Here, we deal with "market-driven" innovation processes that are not brought about by an expressed demand but are prompted by an observation from which the needs are derived, which in turn initiates an innovation program and even some new research. Then, we deal with "market push", a concept that enables the shift from the false opposition of the *technology push* and *market-pull* logics, whose foundations appear to be more in tune with what is currently acknowledged about the emergence and development of corporate innovations [TID 05].

In the example of Nestlé, the consumer's role can also be taken into consideration using a "communicative "logic [BAD 13]. Such an approach has been introduced into R&D operations through the installation of devices, which are dedicated to usage analysis, as shown in the example below with the telecommunications operator Orange.

Orange's digital cafe is a community of more than 6,000 people, outside of the company, interested in digital life and who attend virtual cafes led by a *community manager*. Each year, about 40 digital cafes take place via a Web space, which is unique to Orange.

The objective of this facility is to encourage interactions with consumers or users, the "dreamers", generators of useful information for product and innovation design, with a view to producing new proposals. The work is organized within virtual workshops on themes as diverse as the protection of private life, the child and digital life, the quality of existing digital applications and so on; a workshop can last 5–15 days and requires commitment. Based on the community discussions and an overview of the possible usages, the facilitator proposes the synthesis that allows the production of a knowledge database on anticipated customer expectations that in turn would help to create an immediate offer and test its feasibility.

The digital cafe is part of a Web community "archipelago" created by the enterprise, in which different ideas, product news and experiments (Digital Society Forum, Lab Orange, Imagine.com, etc.) are open to users. The enterprise navigates through this archipelago to create innovative spaces and to identify possible breakthroughs. On the basis of this exercise, the enterprise has formalized a method for streamlining its organic ability to innovate.

The digital cafe is an "immersive, dialogical and experiential" facility for innovating [DAM 14].

Immersive, because it shows that in order to understand customer usage it is not enough to carry out a customer survey; one must know the practices "through and through".

Dialogical, because knowledge about customer uses is based on dialog and interaction.

Experiential, because just as the company learns from the "dreamers", the latter gain skills over time through their participation, thereby becoming competent customers.

The premise of this approach is that innovation does not belong to a function (R&D) and that it must be seen as an extensive, open design process.

Box 1.6. *The digital cafe at Orange*

The example of Orange shows that it has developed a true "communicative model" of innovation based on the intensive use of the Web [BAD 13] in the digital economy sector in particular. Now, "Innovation is the adoption of a new practice by a community [...] Success in innovation intersects the areas of expertise of the innovator powered by the designer, his/her expertise in the field of social interactions and his/her ability to recognize and seize opportunities" [DEM 10]. In this model, remote dispersed communities are demanding a freer organization to propose new innovative combinations, along the lines of what Google promotes via the Internet and social networks, for example.

Innovation is a concept aided by use, which itself is overturned by "social technology" (McKinsey), that is, spontaneity, flexibility, serendipity and adaptation in design, which the user communities know how to display.

1.2.3. *The effects of market globalization*

The globalization movement affects R&D organizations. It triggers a review of the geostrategic and economic approaches of its management models.

Although it has been long affiliated with senior management, R&D has for several years been involved in a decentralization movement: on the one hand, for the operational units that drive specialization in product ranges and markets that require R&D activities linked to their specific strategies and close to the markets they control, and on the other hand, for markets that are geographically diverse due to the internationalization of enterprises, to enable adapting the products to the needs of these markets.

Globalization of an enterprise's strategy is expressed not only by internationalization, that is, the access of a business to markets outside of its country of origin, but it also corresponds to a phase of the enterprise' development, in which the firm's entire value chain, from finance to the distribution of goods and services, is managed at a global level, taking into account the competitive advantages of the location for each element of production.

The internationalization or globalization of R&D of industrial enterprises has been analyzed across the following four phases, which in fact partially overlap [LAP 12]:

– the first phase is one of "low-intensity globalization of R&D" which continued until the 1980s;

– the second phase is one where internationalization of R&D was essentially located in the area of the Triad. This is a concept developed in economic geography by Kenichi Ohmae in 1985, which combined East Asia, Western Europe and North America during the 1980s and the 1990s;

– the third phase followed in which internationalization of R&D resulted in localization in the emerging countries back in the 1990s;

– finally, from 2010 onward, there is a fourth phase that sees the emergence of a reversed innovation process (or reverse innovation) located mainly in emerging countries.

1.2.3.1. *The phase of low-intensity globalization of R&D*

In the competition of industrial enterprises, the traditional strengths of the Western world existed for decades not only in the technological monopoly of major manufacturing processes and procedures but also in the materials and innovative products. These assets that have led companies to focus on products with high added value have demanded a consistent innovation strategy in order to explore eventual breakthrough innovations which allow the creation of products or more competitive processes. These strategies have meant that over the last century, Western companies have dominated in R&D in many sectors (steel, aeronautics, pharmacy, etc.) and in geographically localizing R&D near the decision-making centers of these firms. The main issue of this first phase was to safeguard scientific and technical knowledge as intellectual capital and only transfer product patents and licenses that are sufficiently standardized to foreign countries.

1.2.3.2. *The beginnings of internationalization*

The 1990s marked a major change, which saw the design and production of products with high technological value being localized abroad mainly in the Triad in a transnational approach. During this period, the presence of companies on international markets was no longer limited to product marketing, but included their R&D activities. It then involved, on the one hand, expanding the design processes by leveraging specific areas in the world that had a high scientific concentration such as the clusters located most often in the large Western-based cities. These areas also attracted the main global investment in R&D. On the other hand, it involved being able to break down the design of technological products into independent parts, in particular, in the areas of electronics, biotechnologies and information systems, with the objective of eventually entrusting them to subsidiaries abroad or to partners in these countries. It was during this second period that *open innovation* strategies developed [CHE 03], which highlight the importance for enterprises to build a wealth of expertise from networks to advance innovation via "global networks of open innovation" [SAC 08].

Access to networks of expertise based in overseas territories was also analyzed/considered by Jacquier-Roux and Paraponaris [JAC 11] as an opportunity for the enterprises "to benefit from tacit knowledge that we can only grasp by having a presence and by penetrating networks already in place in the area". This tacit knowledge, which can equate to the intellectual capital of the firm [NON 95], is at the heart of exploratory learning and the processes of innovation. The problem with establishing an R&D center overseas is allowing researchers to participate in networks of co-production of tacit knowledge, such as social networks or institutional research networks, and the co-production of codified knowledge that constitutes patenting.

1.2.3.3. *Localization of R&D in emerging countries*

Since the early 2000s, the localization of R&D activities in emerging countries has been developing. The division of labor between countries that are the intellectual designers and the hard-working manufacturing countries is no longer sustainable. The emerging countries where industry is developing rapidly are investing in their education and research system to become innovative (China, India, etc.)

This third phase of globalization of R&D has the following multiple objectives:

– to take advantage of the scientific and technical development potential observed in several regions and in several sectors (information systems in India, biotechnologies and pharmacy in China);

– to benefit, initially, from a scientific and technical workforce cost well below that in the West;

– to ensure the relay for the development of enterprise activity, particularly since the economic crisis of 2007–2008, where Western growth has slowed considerably, in countries where growth is being maintained with regard to the period at a rate of 5–8% per year;

– finally, to strengthen the design capacity of Western firms with the aim of creating products that are specifically dedicated to emerging countries, as has been observed in the pharmaceutical industry, where large companies such as Novartis or Astra Zeneca have established R&D centers in China first, to advance research on diseases such as gastric cancer or hepatitis, and in India second, to engage in research on tuberculosis.

The Nestlé group has, for its part, established an R&D network everywhere in the world linking the upstream research centers, including that of the headquarters in Vevey, Switzerland, and development centers that are closest to the markets (see Figure 1.1).

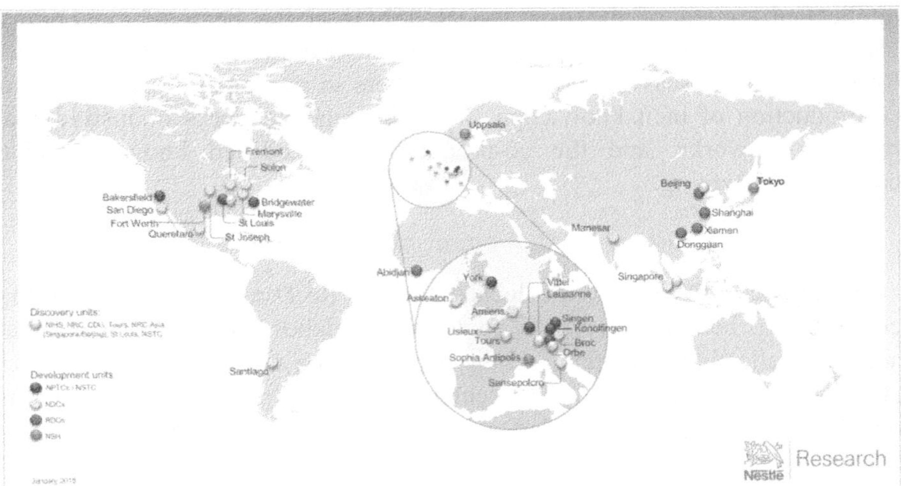

Figure 1.1. *Map of the Nestlé group's R&D center locations. For a color version of this figure, see www.iste.co.uk/gilbert/innovation.zip*

1.2.3.4. *Toward reversed innovation?*

The tendency toward the localization of R&D in several countries through "hubs" is an evolution that has been witnessed in recent times. This is where they connect with different local players that are universities and other local research centers, clusters and the local public authorities, organized in an ecosystem that favors innovation and creates links in the field of innovative exchanges, training or research. The motivation here is multifaceted. There is the economic logic of drawing closer to those markets where economic growth is now taking place. There is also the need for proximity with prestigious academic and research organizations and training institutions to capture innovation and identify the talent. Finally, there is a need for identification of a fabric of small entrepreneurial firms to identify opportunities for technological development or uses to fuel innovative projects.

The concept of "reverse innovation" appears in this period [IMM 09] upon the initiative of General Electric President J Immelt. While he advocated for this strategy from the 1990s onward and successfully implemented it in his own company, he also considered that it is almost inevitable for all the other industrial enterprises. From his point of view, the Western companies will need to develop the entire design process in emerging countries in order to successfully compete with "champion" companies of these countries, which are better placed to propose products, including technological ones, on world markets. The consideration of the constraints of these countries, their limited access to resources and their vast economic needs proposes new avenues for designing new products by capturing needs and local uses and then attempting to develop global products that could potentially be successful on world markets.

Reverse innovation implies a fundamental revision of the design strategy of enterprises in developed countries centered on the research and development of hi-tech products destined for consumers with high purchasing power. In contrast to the traditional design of R&D, it proposes a more frugal technology design and less intensive design engineering [MID 17]. It is a "bottom-up" innovation process.

Reverse innovation is also the mirror image of the strategic management model formulated in the 1990s: "think global, act local" or "glocalization" that wanted enterprises to develop a global R&D strategy aimed at supplying local markets with a holistic engineering design. Since then, companies that rely on reverse innovation have developed the opposite approach: "think local, act global", illustrated by L'Oréal's R&D policy, for example, in which it is no longer about designing global products for global markets, but about starting with product innovations made in emerging countries to make them global and distribute them globally adding to them where necessary through technological improvements that may be necessary in more mature markets.

The traditions of conventional innovation wanted innovation in Western enterprises to be carried out, on the one hand, through top-of-the-range goods, that is, by making more sophisticated products and, on the other hand, according to the capabilities of the technical design of the country of origin and then for it to be launched in the rest of the world. The example of the design of the Kwid, an Indian vehicle from Renault–Nissan, shows a possible split in innovation strategies.

The process implemented by the end of 2011 moves away from usual company practice in vehicle design: to first offer the vehicle on the Indian market, it had to be designed under a strict set of specifications: to drastically reduce costs, they adopted an original design of a small SUV and then modernized it, for example, by installing integrated navigation tools and making it more spacious than the competing models.

The project design story boils down to a permanent struggle by the project leader and his team to "free themselves from the technology, habits and usual management of such a project" [DET 17]. It becomes a reality through strong self-determination in relation to the engineering of the parent company who directed the project team to no longer consider any technical standard issued as secured, for example, questioning the number of securing points on the wheels, reducing the cabling weight, reducing the thickness of the seat runner and a factory without walls or doors.

The project was then controlled by cost targets according to the principles, methods and tools (design to cost), which took into account as closely as possible not only the design choices and customer value objectives but also company costs of working in cooperation with the suppliers. These are up to 97% "Indianized" and offer original solutions even technologically, as collaboration with suppliers started well upstream.

In addition, the project sets up a creative development process where innovation is focused not only on the upstream design phase, but also on the development stages and even the market rollout.

This "fractal innovation" [MID 17] permits an advanced capacity for exploration and learning in the teams and increased responsiveness to sudden changes at all stages of the project. Also, a new method of innovation was developed that neither Renault nor Nissan used; it is a global project based on an innovation approach reversed via the research of frugality in industrial investment.

The result is compelling: investment is three times less compared to the conventional vehicle design and the cost price has been reduced by half compared to a vehicle like the Dacia Sandero in the entry range.

At present, the Kwid has found its market in India (120,000 vehicles sold from 2015 to 2016). It has since been exported from India to South Africa, Sri Lanka and Brazil.

Box 1.7. *The Kwid: Renault–Nissan's reverse innovation (source: Midler* et al. *[MID 17])*

In the example of the Kwid, the R&D know-how from markets other than those of the manufacturer's country of origin (in this case, India) were taken into account, regardless of the innovation practices of the head office and with a global perspective furthermore. This program has also been analyzed as being "symbolic of a return to a transnational model" – like that of the 1990s – as described above [DOZ 12].

1.3. R&D strategies tested overseas: the example of China

As a country where the considerable pace of economic development for more than 20 years has provided significant growth for Western companies, analyzing the R&D strategies that they have developed in China offers a key to additional insight into the challenges that internationalization poses to Western companies. The comparative approach between French and German enterprises in their setting-up strategy in China shows how they are forced to profoundly review their strategy of R&D.

1.3.1. *Western companies' choice to locate their R&D in China*[2]

Initiated in the mid-2000s, the establishment of R&D centers in western China illustrates the shift in R&D strategies around the world and the tensions they create.

Foreign direct investment (FDI) from Western countries has flowed massively into China over the past 20 years. The massive migration of technology and know-how to Chinese companies that are largely controlled by

2 This part is based on a study of French and German companies in China [BOU 11].

the state (which was key to the setting up of Western enterprises) has contributed to "upgrading" the Chinese economy, which has reached a level of technology that allows it to find its own place in the community of groundbreaking scientific nations.

The case of Western companies who have chosen to set up R&D activity in China illustrates the mutations at work and shows their very significant impact on the management of the R&D and expertise there.

1.3.1.1. *Innovation and technology transfer: a different approach to globalization*

Since the end of the 1990s, China has moved gradually away from its image as "the world's first workshop" to establish itself as a power in the area of innovation and research. According to the OECD, Chinese expenditure in R&D increased from 0.6% to 1.98% between 1995 and 2012 (in percentage of GDP), while the number of researchers increased by 77% over the same period. China is far from being behind or inferior to the West in terms of techniques (as we often naively imagine). It has therefore become, according to some observers, an "R&D nation" for foreign economies, which are no longer reluctant to set up their poles of competitive clusters there. For the record, Europe and the United States dedicated 2.07% and 2.79% of their GDP, respectively, in 2012, to R&D expenses.

In the area of R&D, the case analysis of Western enterprises, both French and German, who have a local presence shows that the objectives of setting up in China are not limited to a quest for cost reduction. It is about creating qualified teams for research activities. For this reason, the number of publications, patents and students in China has significantly increased in recent years. It is necessary to be located in growing markets, where new consumer practices are likely to emerge. Finally, it is also useful to take advantage of the "outside critic", who is offset and potentially innovative, which could be provided by a research team working on the other side of the planet. The sectors that invest in the R&D clusters in China are mainly the high-tech ones (telecoms, automotive equipment, transport and energy, etc.). A study conducted by Allouche *et al.* [ALL 08] showed how setting up the R&D activities of companies such as Orange and Schneider Electric was an effective strategy in their approach to innovation. Nevertheless, for most European companies, the strategy has instead been up until now to keep the latest technological developments so as not to surrender them when setting up in China. In fact, in the areas that could be considered as strategic for the Chinese economy (aeronautics, defense, railway transport, etc.), the dominant model for Western enterprises was to transfer the most "underrated" technology to China in order to control innovation and advances in research and development. This strategy has long been preferred by companies such as Alstom, particularly in the 2000s, which, after having established an engineering team in the field of transport in China, quickly removed it due to the fear of providing Chinese teams with the mastery of the new technology developed in the field of rail transport and transport signals: the enterprise's R&D policy was that there should be no center for R&D in China that would work on the core business, a strategic choice that was deeply rooted in the company culture until July 2010. When the transfer was necessary, rules were applied to the developed products for the protection of intellectual property. These were very specific for each party in the joint venture. At the same time, German and Canadian manufacturers Siemens and Bombardier, which have adopted the opposite strategy, that is, the acceptance of R&D technology transfer to the Chinese, gained an enviable position on the Chinese market.

In the field of agri-food, other strategies in the area of R&D could be identified; this is the case of companies like Lesaffre, for which the question of the technology transfer is no longer the real question.

Lesaffre, the world's number one producer of baker's yeast, is a family business originating in the north of France. It has 48 plants in 19 countries with a turnover of 1.2 billion in 2010. It employs 7,000 people. The group, established in China for several years, put in place a very proactive strategy as early as 2008, aimed at becoming the leader on this market and put in place a platform for global export. The group considers that China, which now only represents 5% of its turnover, will constitute with the Asia Pacific 25% of activity before the end of the decade. In fact, for Lesaffre, the question of accepting or rejecting technology transfer no longer exists.

According to its president in China, Lesaffre is already in the second stage of innovation development and technological transfer.

Although fundamental research (which is working to identify new strains of yeast for the whole group) remains in Marcq-en-Baroeul (France), the company has quickly grasped the need to develop in China in order to understand customer needs in closer local markets by engaging commercial and technical teams (with regard to nutrition, the 27 Chinese regions are like different markets). These teams help the company to continue to differentiate itself from its competitors, to develop and push its industrialization know-how and its quality standards while respecting global standards. Lesaffre goes even further: the managers of this medium-sized enterprise are, on account of its turnover figures (MSC), convinced that by being present in China, where there is a decent-sized competitor (Angel), the company also has access to specific knowledge which, combined with its own assets, serves as a technological advantage. This is all the greater because this company, as we have seen, is in close contact with the local environment.

Box 1.8. *The R&D strategy of a medium-sized enterprise: Lesaffre bakery*

Setting up an R&D center in China is not just about technical capacity. This strategy must also consider cultural and human capital dimensions. Like the entities developed by Chinese employees in this context, R&D centers "off-shored" by Western companies are often neglected.

The analyses conducted in several of these R&D centers [LEB 12] reveal a profound gap between the innovation vision of head office and the Chinese vision, which is manifested by a lack of understanding. Therefore, Chinese researchers and engineers at an Orange center for research and technology monitoring complained about the lack of follow-up given to their projects and innovation proposals. The Chinese business models do not have the same conditions for success as the business models in Europe, and it is more difficult for them to convince and to argue with their head office the merits of a research or development project.

In addition, the favored organization of work practices that are based on matrix teams in which the different stakeholders of a project are separated by great distances complicates even more the business processes or even slows down the decision-making process. This is another discrepancy between the understanding of the authority of the parent company/head office and the understanding of the authority in China. For many Chinese designers, research and innovation have become common in Chinese companies and are needed as a free local practice of Western know-how. The Chinese five-year plans 2006–2010 and 2011–2015 for science and technology and innovation development are in this respect an important investment aimed at lifting China to the rank of technological nation.

In German enterprises, the position with regard to technology transfer differs significantly from that of French enterprises. Despite the very real risk of illegal transfers, Germany has relied on a relatively early scientific and technological cooperation with China, embodied by the signing, in 1978, of a bilateral government agreement. The following year, China and Germany decided to discuss the German standardization (DIN): the diplomatic business trips of technical specialists and economic representatives increases significantly, such that over time, many of the standards of the German model have been adopted in China. A new method of cooperation on projects is put in place, followed soon after by an institutional cooperation that leads to the creation of centers of bi-national research in 2004 and 2005. The Federal Ministry of Education and Research worked in partnership with the Chinese Ministry of Science and Technology as well as with the Chinese Ministry of Education.

In addition to intergovernmental scientific collaboration, German private not-for-profit research organizations developed a direct cooperation with their Chinese partners; therefore, the Sino-German Center for the promotion of research (Deutsch-Chinesisches Zentrum für Wissenschaftsförderung) managed jointly by the German Research Foundation (Deutsche Forschungsgemeinschaft, GFR) and the National Natural Science Foundation of China came into being in Beijing. The Fraunhofer Institutes have furthermore opened two research centers in partnership with the High Technology Research and Development Center of the University of Astronautics and Aeronautics in Peking: the Sino-German Joint Software Institute (JSI) in Peking and the Sino-German Mobile Communication Institute (MCI) in Berlin. The Max Planck Institute created in turn, in 2005, a Center for Computational Biology in Shanghai, in collaboration with the Chinese Academy of Sciences. The Leibniz Community have been working since 2004 on setting up a network linking its own institutes, the research institutes in China and its industrial partners, in order to be able to separate, distinguish and test biologically active substances coming from plants from traditional Chinese medicine. Finally, the Helmholtz centers work in cooperation with an extended network of major Chinese institutions, especially in matters of the environment, health and energy.

At present, many German companies view China as a market of strategic importance. According to these companies, the evolution of the European economy will depend on the success of positioning in this market, among other elements. This is particularly true in light of the very rapid technological progress made by China, which results from a proper innovation effort or a technology transfer taking place. The presence of German companies on the Chinese market is therefore crucial to preserving German and European competitiveness as well as to developing joint standards.

Consequently, for the German companies, technology transfer should not be a "problem" because "the only risk that can threaten the competitiveness and innovation of the German economy lies in a possible shortfall in the domestic innovation and research policies. The fact that China launched a set of programs to rise to the rank of technological nation then became widely known. However, evaluations of their true progress differ. For Germany, this can only be an incentive to preserve their competitive advantage and our capacity for innovation on two levels: due to the proactive attitude of the businesses involved and through the creation of a supportive political environment by the state. However, under no circumstances can Germany rest on its laurels – "the competition never sleeps" [STÄ 10].

Therefore, we can see that European, French or German companies planned the transfer of Western technology with a great deal of delicacy: ensuring, for example, the fragmentation of know-how in the organization and processes, including the management of IT, knowing how to identify the main risk sources and always keeping control of key skills.

1.4. Conclusion

The mutual effects of intensive innovation and globalization allowed new players to enter the scene of world competition, which in turn created new challenges for businesses.

Pushed to develop their activities outside of their historical markets, Western companies have also been led to internationalize their R&D activity. Analysis of their practices in this regard shows that their R&D strategy depends significantly on the exposed or unexposed nature of their competitive positioning.

If the majority of R&D strategies align with the need for rapid innovation expansion, then there are several factors that must be considered in order to analyze the differences in the practices of the enterprises studied. First, it is the major role of innovation that leads some to develop global products for profiting on very large markets like telecommunications and mobile phones, while others try to grasp the value created by innovation in emerging markets. The growing importance of the client, already present in other business activities like marketing, quality control and sales, is now also present in R&D, where it becomes an integral part of the innovation process with a logic of co-design or collaborative design. Finally, there are practices of localization of R&D in BRICS, for example, to take advantage of the newly emerged "technological nations".

The example of China and its attractiveness for enterprises in the sphere of R&D management reflects the tension between the need to capture knowledge and talent in this new high-tech champion nation and the desire to protect the innovation capital that has been accumulated.

2

Work in R&D and its Transformations

Work in R&D has changed considerably since the 1990s, and currently, it is still going through important changes, some of which are only in their infancy. From the past, present and future transformations of R&D efforts, strong challenges are emerging in R&D management, from the management of project activities to human resources and knowledge management (see also Chapters 4–6). To understand the way in which R&D work has transformed, in section 2.1, we describe its specific characteristics in relation to the work carried out in other areas of the business, while also considering the heterogeneity that exists within R&D activities. In section 2.2, we explain the main developments in R&D over the last 25 years, which are currently well established in the organization of R&D in large industrial enterprises. We begin section 2.3 by highlighting how some of these mutations in R&D can give rise to tensions, for example, the rise of the project form and increasing performance pressures. Then, we indicate some current trends, which are still largely in the making, as well as some concerns regarding the future of R&D work and function, notably the questions of outsourcing, internationalization of R&D and the effects of open innovation and digital revolution of R&D work.

2.1. Specifics of R&D work and its heterogeneity

R&D comprises specific activities, which stand out from other major functions of the company mainly because of their unconventional nature. Furthermore, here we also intend to demonstrate that R&D is not a homogeneous entity. In fact, the work varies strongly between the upstream and downstream parts of R&D, depending on whether the researchers and engineers in R&D or the technicians are concerned. It also depends on the business sector (agri-food, aeronautics and pharmaceuticals, etc.), the scientific and technical field (chemistry, biotechnology, informatics, mechanical, etc.) or company strategy in innovation or even its size. In a contingent approach, knowledge work carried out within an R&D business – with its specific features compared to other functions as well as its internal "variability "– is essential in order to be able to design and implement relevant management practices, given the issues that R&D has to face. However, Katz [KAT 04] mentions how knowledge and practices in management were developed first on more operational company activities, especially production. This tropism makes it difficult for managers from an R&D professional background to grasp its distinctive nature and implement managerial practices adapted to the challenges and logics that are specific to R&D.

2.1.1. *Non-routine and knowledge intensive work*

R&D activities are non-routine and unusual in that they are aimed at building new knowledge and artifacts. They differ from the activities performed by other functions of the enterprise, which are more stable or cyclical, more predictable and therefore more pre-determinable and manageable. This uniqueness is a common trait for R&D business, even beyond the differences between the "R" and "D "works.

The Frascati Manual of the OECD [OEC 02, p. 34] specifies that the term R&D – creative work undertaken systematically to increase the knowledge base for developing new applications – refers to three activities, namely basic research, applied research and experimental development. Basic research is experimental or theoretical work undertaken primarily to acquire new knowledge of the underlying foundation of phenomena and observable facts, without any particular application or use. Applied research also involves original work undertaken with a view to acquiring new knowledge. It is, however, directed primarily toward a specific practical aim or objective. Experimental development is systematic work based on existing knowledge gained through research and/or practical experience, in order to begin the manufacturing of new materials, products or appliances; to establish new techniques, systems and services or to improve the already existing ones.

While activity differs on whether the upstream part of R&D (the "R" geared toward exploration and ideation) or the downstream part (the "D" geared toward the development and implementation of innovations) is considered, the common dimension is the work on the new ideas and artifacts.

In fact, the mission of the R, as with the D, stops with the invention and demonstration of its functionality at a technical level. Then, other professions in the company take over the management of day-to-day operations that arise from the development of product or process innovation.

Another characteristic of R&D business is its fundamentally uncertain nature. This uncertainty is linked to the quest for novelty, and it is even more powerful when the upstream exploration activities are considered. In research, unpredictability is considerable, one never knows *ex ante* what we will find exactly, or at what time horizon, or even sometimes if we are going to find anything, never mind knowing what volume we are going to sell and at what price, etc. The concept of serendipity emphasizes the fact that we may end up with unexpected results in comparison to initial objectives. In the field of development, the degrees of uncertainty are lower on technical and economic levels, as well as in terms of feasibility and value. Therefore, Le Masson *et al.* [LEM 06] define development as “a controlled process which activates skills and existing knowledge to specify a system (product, process or organization) which should respond to well-defined criteria (quality, cost, deadlines) and whose value has already been clearly conceptualized or even evaluated”. However, unforeseen difficulties may occur and alter the timescale, costs and specifics of the object under development. Across the business sectors, failure rates in R&D can be very high (see also Chapter 6); therefore, in pharmacy, more than 95% of projects are abandoned, considering that a laboratory launches a very large number of small competitive projects, but with a low rate of success. In contrast, a car manufacturer only simultaneously accepts a limited number of new vehicle design projects; however, the economic stakes of each are very high. Uncertainty is weaker here; projects generally convert to business when they go to market, all the more so in cases where the projects form part of a product line [LEM 06], for example, Renault with the successive models of Clio.

Depending on the industry, R&D encompasses wide-ranging scenarios. Companies 'strategies in terms of innovation (see Chapter 1) also heavily influence the R&D activities that are undertaken within. Therefore, in some cases, the R&D function consists essentially of development activities because of the low-tech nature of the sector and/or because the enterprise strategy pivots on incremental innovation, in an operational logic (March 1991). In other cases, in hi-tech sectors and when the company strategy targets the introduction of break-through innovations, research activities will be also present, with varying importance, alongside development activities, in order to support the exploration logic (March 1991). We must consider therefore the distribution of tasks between R and D, beyond the clusters that globally quantify overall input into R&D (expenditure as an absolute amount or as a percentage of turnover, workforce in R&D and percentage of the global workforce).

In the enterprises that actually perform the work of R and D, different structural choices have been made: in some, R and D are carried out by the same professionals who share their time between these different activities; in others, R and D are quite separate. The question which arises here and which is not clear-cut is how to succeed in being ambidextrous [TUS 96], i.e. how to jointly meet the requirements of operational innovation based on the efficient use of technological and marketing skills existing in the company and the requirements of exploratory innovation that requires the renewal of these skills on one or another of these dimensions. Organizational ambidexterity is when the development and exploration activities are separated, and contextual ambidexterity [GIB 04] is when they are carried out within the same plants by the same people (see Box 2.1 for an illustration).

STMicroelectronics opted for a structural ambidexterity. The HR manager points out the existence of divisions of professional researchers who do not move from their structure. Therefore, the group has created an entity called AST ("advanced system technology") composed of several research laboratories in the world and comprising approximately 250 researchers, which has the mission of exploring new systems and structures for a 3–5-year term. This organization carries out many new knowledge exploration projects and is evaluated on its ability to identify new innovation leads.

Saint-Gobain, on the other hand, has organized its R&D on the principle of contextual ambidexterity. We quote the Human Resources Director of Saint-Gobain Research: "Personally, I do not believe much in the idea of only having researchers who work on exploratory subjects, completely disconnected from the rest [...] At Saint-Gobain, we have never sought to recruit some individuals for exploration and others for exploitation, precisely because we operate jointly". Therefore, Saint-Gobain Research, which is the enterprise's central process laboratory, consisting of 305 employees, is entrusted with exploration assignments, anticipating the future needs of the group and working on breakthrough projects and with operational assignments in problem solving mode in mature projects, where the objective is to improve existing products.

Box 2.1. *Should R and D be together in the same unit or separated? (source: Dhifallah* et al. *[DHI 08, pp. 167–168])*

The definition of exploration innovations (above) emphasizes the fact that they rely on a renewal of skills in the areas of technology and/or marketing. While operational innovations are part of a greater continuity, in a context of incremental improvement compared to the existing one, they are also based on highly developed knowledge and expertise. This is another characteristic of R&D – to be knowledge-intensive: knowledge being the raw material to R&D. Knowledge is the essential input to research activities, which aims to exceed the current state of knowledge (see the above definition of R). Knowledge – particularly but not only the one produced by the R in-house – is also critical for the D (see the definition by the Frascati Manual). Knowledge is also the output of R. With regard to the D, even though it is not its primary mission, it produces knowledge due to the development of new artifacts. R&D professionals are archetypes of knowledge workers who have become essential in our knowledge-based economies.

Therefore, R&D acquires knowledge generated within the organization (by the R, by the D) and from outside (from other companies and through the higher education and research system, or possibly from individuals – customers, fans, etc.) and creates new knowledge. By doing so, it makes current knowledge obsolete (even that mastered by the members of its own company). R&D is engaged in a dynamic of "creative destruction " aimed at knowledge and skills as well as products and enterprises. While this term has been brought to the foreground by Schumpeter [SCH 42] making innovation the driving force of economic growth in the long term in capitalist economies, recent times have been marked by a phenomenal acceleration. This acceleration is visible in all countries and in all sectors, and it is linked to scientific and technological progress at the global level as well as to the prevalence of competition through innovation and strategies for the accelerated obsolescence of products. However, the intensity of these processes varies according to the sectors. Thus, in certain markets and/or in some developed scientific and technical areas, developments are weaker, slower and more predictable. The dynamics of know-how are less turbulent in the areas of heavy chemicals, agri-food and the construction industries than in the fields of nanotechnology, biotechnology and information and communication technology.

This context of acceleration in the dynamics of the creative destruction of knowledge generates a risk of obsolescence for individuals, groups and companies. To overcome this, it is essential to continuously advance knowledge and skills and to reposition areas of expertise, in conjunction with the specific dynamics of both the scientific and technical fields and those of the markets and the strategies of the enterprises. Chapters 4 and 5 will explore the issues of continuous learning, by investigating, respectively, the role of HRM – with regard to individual learning – and collective knowledge management. Box 2.2 shows the issues at stake regarding learning issues such as those indicated by the R&D engineers of a technology SME in the medical devices sector.

Learning capacity is a key criterion for selection in recruitment: "We look, especially, to see if they are competent and if they know how to adapt to the different codes. We use a lot of coding language and we work on Linux. A new employee must adapt very quickly and we try to pick that up in the interview [...] There must be two things: the spark and the ability to adapt. The spark because they must want to learn. And then we try to see if they can adapt and fend for themself."

The need for training is ongoing and it happens through self-training practices: "The first quality they ask for in the department is autonomy, since there are so many of us. It is a kind of permanent ongoing training. When you arrive, you have to learn on your own. But ok, we are selected on our ability to be effective and autonomous. In our department, it takes 1 year and a half for a person to bring significant added value."

And others offer their opinions:

– "The problem is that there is only one engineering school in France that teaches our core business. Therefore, you learn on the job. When I first started my internship, I started to read, to work in the lab, to do experiments..."

– "There is a large lab in France that works on the same things as us. We get researchers from this lab to come to us on internships. I think that we manage to master all the new things that come out of this lab."

– "The thing is that being on a market with strong competition, you have to go and look for ideas somewhere else, assimilate them, to then develop something new."

In keeping with the needs of the company, the employees can be called upon to spend time in new scientific and technical fields: "I was not an expert in the mechanical field, but I had to work in this area. It gave me the chance to do something different and today I understand how it works."

Box 2.2. *Work in R&D: continuous learning (source: Gledel [GLE 16])*

The constant knowledge and skills dynamic that R&D individuals and groups must master is also linked with innovation in the field of scientific tools. In this way, R&D tools have evolved, if we consider the progress of usage in the area of calculation (with the advent of computers and their democratization), observation (with the electronic microscope), experimentation, measurement and profiling (with the mass spectrometer), etc. Some disciplines have gone back already several decades to the "Big Science" era, resorting to very expensive tools: astrophysics, biology with genetic sequencing and biotechnology, physics with particle accelerators, etc.

These technological developments in terms of work tools mean that monitoring needs to take place first of all in order to guide choices of investments that are becoming heavier, as well as ongoing training at the individual and collective level to be able to control the use of these new tools and follow their successive improvements. If these tools enable us to gain in efficiency (work more quickly) and effectiveness (doing things better), they take part in a competition where they must possess better equipment than their competitors.

As for the use of these tools at the training level, some of which are extremely complex and developed by a very small number of suppliers, they play an essential role in the provision of specific training courses on the instruments and equipment that they sell to their customers. If we consider the large, even the very large, instruments (particle accelerators, etc.), due to their cost and the need for engineers and technicians for their maintenance and operation, industrial enterprises sometimes choose to use publicly funded equipment or to co-finance with other private and public participants on technological platforms that then benefit multiple users.

2.1.2. *The work in R&D: between interactions and engagements with the surrounding environment*

The previous section explained how R&D is fundamentally open to its environment, and particularly to its scientific and technical environment. R&D professionals have the important job of external monitoring, essential for R&D and the innovation capacity of the enterprise. In addition, their activity generates knowledge that is – partly – "redistributed", diffused, evaluated, challenged and reused by actors in this external scientific and technical environment. It is impossible to work in one's own area and ignore the knowledge and technology developed by others worldwide. In addition, a part of what R&D produces is aimed at external diffusion (patents, publications, building up a reputation, influence on standardization, etc.), and this is one of the main R&D missions, apart from contributing more directly to the enterprise's innovation processes.

This opening to the outside is not recent: it is con-substantial to the R&D business and indispensable. The R&D professionals in the enterprises have always devoted part of their working time to reading published works, analyzing patents and exchanging with other scientists – industrial as well as academic – within the framework of scientific and technical conferences. These links with the external scientific and technical environment take place at different scales: local, national, regional and international. However, scientists have not waited for the economic globalization of the past 40 years to exchange views on a wide scale within international networks of scholars. History of science illustrates the importance of these exchanges as well as the early globalized nature of competition in the field.

The links with the scientific environment do not only come from interactions but also take the form of genuine collaborations. Furthermore, the R&D professionals are not content to just read the work of other scientists or to exchange knowledge at conferences, but they work together. Well before the recent success of the concept of open innovation (see Chapter 1 and section 2.3.2 of this chapter), multiple forms of cooperation already existed with academic research. Box 2.3 provides an illustration.

Rhône-Poulenc, which was the main pharmacy and chemistry enterprise in France before its division in 1998 between Aventis for the pharmacy part and Rhodia for chemistry, has always maintained a variety of strong links with higher education and research. Some of its major innovations, from the beginning of the 20th Century, emanate from collaborations with researchers from the Institut Pasteur. In the 1970s, Rhône-Poulenc was the first company to sign a framework convention with the CNRS, proof of the importance of their relations: collaborative projects, hosting of doctoral students (funded through grants from the company's own finances) and post-doctoral students, movement of researchers between the enterprise and the academic world, participation in the Steering Committee of the R&D of eminent researchers such as Pierre-Gilles de Gennes, involvement of academic researchers in advisory roles for R&D teams, etc. It was with Rhone-Poulenc that the CNRS created the first UMR with the industry (joint laboratories merging personnel from CNRS and Rhône-Poulenc, with co-financing) as well as the first international UMR (first in the United States, and then in 2011 in China).

Box 2.3. *Rhône-Poulenc and then Rhodia: close links with the academic world (source: Gastaldi [GAS 07] and CNRS website)*

These strong links with higher education and research contribute to a strong porosity between these professional worlds. Industrial and academic researchers share – in addition to identical training and the same socialization process – similar professional identities (this is less true of D engineers) (see also Chapter 4). They form epistemic communities around scientific sub-areas that cross organizational, institutional and geographical boundaries. The engineers of D are stakeholders of external technical communities, less academic and geared more toward the practice in which they share ideas with other professionals in their field.

Thus, beyond just academic researchers, the R&D professionals also maintain relations with other enterprises (suppliers, customers, competitors, partners), in particular with their R&D teams. For example, researchers are in relation with the suppliers of materials and scientific instruments, and development/design engineers with the suppliers of spare parts.

R&D professionals can have relations with the regulatory agencies. In the telecom sector, for example, it is essential to ensure interoperability between countries, operators, hardware vendors, etc., and it is the role of the European and global standardization organizations to determine the technical standards to permit this interoperability. In a survey conducted on the R&D professionals at Orange, it is clear that with the presence of standardization bodies and the technical lobbying that is done, there is a role granted to R&D engineers. Indeed, company representatives must be sufficiently sharp with regards to technical skills to be able to understand the issues in the negotiations in order to try to influence decision-making and settlements to be reached. Thus, R&D designers work alongside public bodies and competing companies.

2.1.3. *A job characterized by a certain degree of autonomy and occupational regulations*

The work of R&D is also distinguished by a significant degree of autonomy, which stems in part from its other traits: an unconventional job, one geared toward the creation of knowledge and/or new artifacts; a knowledge-intensive occupation requiring highly qualified people who are experts in their field and workers who participate in knowledge communities that cross the boundaries of the organization. Researchers, and to a lesser extent, R&D designers, are "professionals" within the American sociological meaning of professions [CHA 12]. They are characterized by their expertise, as already discussed, as well as by their autonomy, their involvement in the work and the profession, their identification with a group of professionals and their ethic and cooperative efforts in the establishment and control of professional standards (see Chapter 4).

On account of both their professional identity, of which autonomy is an essential element, and the characteristics of their business, R&D professionals are much more autonomous than the other professions in the company, apart from other workers in creative industries (design, advertising) or managers at a certain level.

However, differences exist between R and D. In research, autonomy can bear on both the objectives of the work (although in essence nobody can define *ex ante* what is going to be and what should be found) and the organization of work (how does one define a process that leads to an undefined objective?). In addition, professional regulations are particularly needed in the world of research, with heavy peer pressure (inside the company, but also sometimes externally) in the performance evaluation of industrial researchers at different times of their career (selection, work and skills appraisals, promotions, etc.). In the area of development, the pressure in professional, managerial and organizational schemes is different. Thus, during the development of a new product, the objectives, as well as the job timescale, are often within a precise framework of specifications in terms of quality, cost and deadline, and if the job cannot be completely controlled, it is still closely monitored by the manager.

In section 2.2, we will emphasize the reduction in R&D autonomy in recent times, with a stronger managerial intervention in this function than before. However, R&D still continues today to benefit from a greater autonomy than the other functions of the company (which are also the object of significant streamlining), and this autonomy remains greater in R than in D.

The companies have indeed acknowledged the need to grant a certain degree of autonomy to R&D professionals, which leads to specific management practices. Thus, Google gives 20% "free" time to its R&D engineers so that they can work on what they want, outside of the objectives that are set by the organization and outside of the projects to which they are assigned. This is done as much in a spirit of reward, thus responding to the aspirations of these professionals in terms of autonomy, as in a logic of managing innovative activities. Indeed, it is impossible to explore new questions, have innovative ideas, break free from existing thought processes and push through the boundaries of what we know and what is possible in work frameworks, which are precisely stipulated in advance, without the possibility of making adjustments along the way and under strict control and the scrutiny of a third party. Google's practice of "free" time cuts the R&D engineers some "slack", so that they can be creative (see Chapter 3 on these issues). Galindo [GAL 17] gives another example of more autonomous managerial practices in the case of Withings, a start-up (now acquired by Nokia) that markets Internet-ready devices for the health sector. Entrepreneurship – as a lever of innovation – is encouraged, and this is done by giving employees scope to choose their activity. One developer explains: "At Withings, they ask you: 'What would you like to do? Would you like to work, on this, this or this,' so that's interesting." [GAL 17, p. 62]. The table below also illustrates the importance of autonomy in R&D from the viewpoint of R&D engineers. The quotes illustrate how autonomy does not mean absolute freedom and how the existence of frameworks and a certain degree of constraint can also be stimulating in terms of creativity and innovation.

The following quotes are extracts from the responses from the R&D engineers of a technological SME in the medical devices sector who were interviewed on the key factors promoting innovation:

– "Freedom is key: freedom to choose tasks, freedom in the organization of your days and in the organization of your work each week. It is creative work, so you must be free. Too many routine tasks kill innovation. If you have your head in the game too often, it is impossible to innovate […] You need time, a time slot, to do something other than what management requests of you, to devote yourself to reflection and research. We thought about doing Research Fridays, but unfortunately with new product release outputs we have not managed to do it."

– "Yes, autonomy is good but sometimes we need a framework. If there are no safeguards in place for the people who get ahead of themselves, they may not progress as they should in the end. We have experts with strong personalities; this is good for creativity. But we do not need to be super innovative all the time. Without some consistency, we will become deluded. Everything will be good, but not enough to sell to and satisfy the client. It is the coordination of skills that will be important."

Box 2.4. *Autonomy – as a condition of innovation (source: Florian Gledel, M2 HR, Aix-Marseille University, Dissertation, 2016)*

Noting this need for a certain degree of autonomy in work, management practices developed in R&D are moving away from the "classic model ." Managers act more like coaches than hierarchical superiors exercising a permanent and direct control on work and behavior. Beyond the impossibility of carrying out a detailed specification of the work (as the production or administrative departments may do), the position of the manager is also complicated by the existence of frequent hyper-specialization in R&D, which can make them less competent, or at least less of an expert, in the knowledge areas his colleagues are working in. Work organization falls more within the organic model than the mechanistic model [BUR 61]: formalization reduced, wide spectrum of activity, multiple coordination and decentralization of power. Coordination comes predominantly from mutual adjustment (by direct contact, without going through hierarchical intervention) in organizational forms, which are similar to Mintzberg's adhocracy [MIN 82]: a flexible type of organization that adapts according to the needs and constraints of the business activities to be achieved and complex activities, which are part of dynamic environments.

Earlier comments about professional identity and management methods relate to researchers and engineers in R&D. The technicians are, for their part, in a less favorable position in terms of autonomy, given their support role compared to researchers and engineers who have the power to define their work, with respect to its purposes and objectives, but also its content, organization and timing. The Frascati Manual [OEC 02] defines the researchers and engineers in R&D as "specialists working in the design or creation of new knowledge, products, processes, methods and systems, and in project management", and technicians as "people whose main tasks require knowledge and technical experience in one or more fields of engineering, physical and life sciences or social sciences and humanities. They are involved in R&D by carrying out scientific and technical tasks by implementing operational principles and methods under the supervision and control of researchers". While the difference between technicians and research engineers is significant, the differences between the research work undertaken by researchers and research technicians and the work shared between engineers and technicians are also considerable.

Thus, in section 2.1, we pointed out both the specific features of R&D work compared to other functions within the enterprise and its non-homogeneous character. Therefore, we could say that R&D is contingent to a large number of aspects such as personnel categories, the size of the organization, its strategy, its business sector, the scientific and technical fields, depending on whether the activities are upstream or downstream, the degree of autonomy, the nature of the project, its timescale, how open it is to the environment and the kind of external partners it has, the speed of knowledge advancement as well as its links with materiality (see Box 2.5).

Depending on the business sectors of the enterprises and the scientific and technical domains, the relation with materiality in R&D work can be very different. This does not necessarily cover the gap between R and D, because an experimental project in chemistry, on the lab workbench, handling many substances and many scientific instruments (test tubes, pipettes, measuring devices, centrifuges, scales, etc.) is grounded more in the material than the development of software, for example. Having said that, considering the world of chemistry, if the material dimension is very present in research (and while the results of a research project can only be recorded in written form: lab notebooks, formulas, etc.), it has a deeper meaning when it is an experiment of an industrial process used to manufacture a chemical molecule, even deeper when it comes to the engineering process stages of plant manufacturing, which are incredibly complex.

We also see how in chemistry we talk of research, and particularly in the synthetic industry, about research centers, pilots, and about engineering process, whereas in the automobile or aeronautical industry, we speak of engineering consultancies, techno-centers, prototypes and templates.

When, depending on the industry, the result of the innovation process is a plane, a car, a yogurt, a software, a protein, a grain of silica, etc., then the relation to the material is inevitably different. In the upstream phases, the use of industrial design or the prototype varies significantly according to the field, whereas in chemistry, a design can complement a chemical formula, and in manufacturing industries, tangible goods designs are essential tools, all the more so if they are composite goods like a car, which has more than 10,000 parts. In the area of information and communication technology, including the services area, design is very important, including questions on user-friendliness and cognitive ergonomy.

Box 2.5. *A different connection with materiality depending on the scientific domains. Image source: Crystal Kwok, unsplash.com*

After having pointed out the special features of R&D work, we must highlight how it has evolved since the end of the 1980s, with an important acceleration over the past 20 years. In section 2.2, we focus on describing the main transformations, which are currently rooted in R&D work. In section 2.3, we demonstrate, first, how some of them are a source of difficulty and, second, how R&D work is really called into question by more recent developments including those that are still to come, as well as internationalization, open innovation or digital revolution.

2.2. The main transformations of R&D work since 1990

The R&D function has transformed over the last 25 years with greater transformations occurring in the development function compared to research. According to the relevant countries, the timescale and the form and intensity of these mutations vary, so the United States, for example, has experienced these R&D advancements well before France. The emphasis is successively placed on the advent of project management organization, the strong interaction of R&D with other departments, greater dependency on other departments and the managerialization of R&D executives 'work.

2.2.1. *The advent of project management and of the concurrent engineering model*

The work of R&D has evolved considerably with the introduction of project management [MID 93a, MID 93b, GAR 03c], which has gradually asserted itself as the model for organizing design processes for new products. Project organization is not new (if we think about engineering and construction – bridges, dams, etc. – nuclear power, space), but what we characterize as the project management today refers mainly to the organization and design of activities according to the principles and the model of concurrent engineering. In this, a transversal project combines and coordinates the actors of different departments that are involved in new product development. These actors will contribute to the project, moving forward together toward the same unique goal, exploring and engaging in parallel work on the different aspects of the project (product design and production process design; technical aspects and functionalities etc.), which, to date, have been dealt with in a successive manner. It is also about regrouping all the different units involved as well as anticipating as far as possible certain responsibilities and decisions. The idea is to provide a degree of freedom in the upstream phases and to postpone tasks and decisions, which require the commitment of heavy and strategic resources as late as possible once knowledge has been gathered, thus limiting uncertainty. The aim is to complete the project as quickly as possible while reducing costs.

It is useful to refer to existing organizational structures [GAR 03a, GAR 03c] to understand the logic and contributions of concurrent engineering. Earlier, the industry was organized following a Taylorian model in operation, working sequentially and completely disjointedly from the other units from upstream to downstream in a science pushing logic. Takeuchi and Nonaka [TAK 86] used the metaphor of the relay race to reflect this organization method: research begins, passes over to development if it has found interesting results, and development works alone to pass over to the industrialization and then to marketing, etc. There is no interaction, or collaboration between the different departments that work successively within the strict boundaries of their expertise, seeking to optimize the quality of the work done in the sequence of the innovation process and/or on the part of the product for which they are responsible.

Such a sequential Taylorian model can be inefficient and ineffective in many ways. First, none of the actors are responsible for the overall product development, with each one working in turn and under the authority of their department or business unit. The duration of the project therefore is the total length of each of the processes consecutively, each one considering that the importance of their work deserves a substantial amount of time, which then results in the overall time being exceeded. In addition, the success of an innovation process is conditioned by the overall product quality, by the capacity to satisfy consumer needs and by the acceptable price that consumers will be willing to pay for such a product (taking into account the use value of the product and of competing products on the market), and not the optimization of each of its dimensions. It is typical to see R&D designers, for example, make their inventions more sophisticated, without considering the customer usage, in turn risking an excess in the cost and the sale price. Going further, it is very complicated for upstream actors to integrate the issues and constraints of industrialization, purchasing or marketing and sales, while they are never in contact with them. They have to imagine what could be of interest to a customer whom they do not know, which is not easy. There is a lot of waste then, and R&D projects that have consumed time and resources are then stopped in the downstream phases because they are finally deemed irrelevant from the viewpoints of clients, markets or company strategy or even deemed too expensive. In contrast, the downstream actors may encounter difficulties in the industrialization of the product or have to respond to customer dissatisfaction, without managing the underlying processes. Then, it is difficult to resolve these problems involving long cycles, especially since the R&D engineers who were at the origin of the product development have since moved to other projects or have even left R&D. This causes delays and significant costs, which can be fatal for the outcome of the innovation process.

While the competitive context is evolving in the direction of increased competition, with a new rhythm of innovation and different time management, the pitfalls of conventional design organization methods become unsustainable and prohibiting vis-à-vis competitors. The analysis of the virtuous practices of Japanese companies in design leads to the United States and Europe testing a new model called concurrent engineering. It involves the creation of temporal structure (horizontally organized project structure), which crosses over and draws on the permanent resources of the company (vertical axis by functions and business units), thus generating a provisional matrix structure. Overall objectives are assigned to the project (and no longer to each business unit), and it is the responsibility of the project team to meet the targets. Each project is given a framework, a defined cost, quality (with technical and functional specifications to attain) and a timescale. Project control is no longer performed by the business units. Instead, a steering committee specifically defined for the project will review the project at each stage. A person will generally be designated as project manager, responsible for its progress and compliance with the cost-quality-timescale specifications. Depending on the configuration, the level of authority will vary, but this tends to increase in correlation with the size and the strategic importance of the project (see Box 2.6).

Renault from the end of the 1980s has adopted new design organization methods, and in every project:

– a project manager is responsible for the entire project. They are appointed by the directors (who give the project manager strong authority) to whom they are accountable in terms of budget, schedule and quality of the vehicle;

– project leaders represent the different technical departments (business units) on the project: research, industrialization, procurement, product, design, logistics, planning, quality, etc.;

– as for the departments (or vertical structures), they offer their expertise and skills through the allocation of staff on the project.

Over time, and in correlation with the strategic importance of the project design of a new vehicle, the project managers have increased authority. For example, they initially had to negotiate with the department directors to obtain resources. With the change, the general directors assign them their own budget, as well as decision-making rights similar to the directors of the business units. In the same way, they have acquired a power of co-evaluation, alongside the department directors and the actors from the business units involved in the project.

General management has established rules common to all projects, which only they can modify. The following three guiding principles underpin these rules:

– solidarity between business units with the commitment to meet the objectives negotiated between the project directors and the business unit directors;

– accountability and delegation to the trade representatives of each department to maintain coordination within the project; and

– transparency and the recognition of a right to error to enable identifying and resolving problems.

Box 2.6. *The organization by project at Renault (source: Zannad [ZAN 08])*

Project teams bring together from the beginning all the business units concerned, and in particular, the downstream actors are integrated very early in the project, to incorporate the constraints and challenges of industrialization, supply, marketing, etc. In contrast, the upstream actors now stay with the project through to completion. The rugby team metaphor could be used here, as intensive communication and a strong ability to cooperate are essential. Charue-Duboc [CHA 97] indicates the principles that make concurrent engineering effective: global optimization all around the project and not only on one of its dimensions; anticipation of development problems; responsiveness (the reaction speed to the uncertainties encountered and quick resolution of problems) and client orientation.

Beyond these generic principles, projects will cover very different situations, according to the industries. Therefore, in the car industry, each project is very important and consequently project participants are very often dedicated entirely to the project, in "out-mode", i.e., in working outside their skill set. The project team is localized on a "plateau", which is a physical space where all individuals within the different business units contributing to the project meet, in order to facilitate the coordination, cooperation, through mutual adjustment and close connections that develop over time. This may encourage issues and pressures from different business units and from the different dimensions of a vehicle at any time (product process, technical marketing, quality cost, interactions between the different sub-systems: motorization, steering, braking, electricity, etc.). The suppliers are sometimes included in these projects to anticipate problems of component integration in the new model of vehicle or even for co-design of new parts or sub-systems of the vehicle.

In the chemical industry, there are more projects but smaller ones, at least in the upstream phases, where the failure rate is high. Therefore, a project that "works" becomes much bigger, especially if it is accompanied by a development project of a specific production unit. At Rhodia, the researchers we met were working on several projects at the same time. They were in frequent contact, remotely or through meetings, with the other business units of the company (downstream business units of R&D, process engineering, internal customers, technical marketing, etc.), but they stayed in their research laboratory, where they were undertaking tasks necessary for the progress of the various projects that had been assigned to them [GAS 07].

The term project also covers a wide variety of situations, particularly according to the business sectors, with regard to their strategic importance, size, timescale, budget, complexity, the customer (i.e. if it is a project carried out for themselves or for third parties, an identified third party or a potential third party, industrial customer or end user) or even their style of leading and their principles for managing (see Chapter 6).

Apart from the wide variety of projects, it is still possible to point out a few features of R&D projects (see Table 2.1).

Skilled work	Project work
Working in the framework of a traditional functional structure, under the authority of the business unit director	Working in a matrix structure under the dual authority of the business manager and the project director/manager
Homogeneous and unbounded timescale	Heterogeneous timescale [MID 93b]: time sometimes passes very slowly, sometimes very quickly, with benchmarks Timescale bounded by the project deadline
Continuity logic: the business units are focused on ongoing renewal and deepening of knowledge and skills in their scientific or technical domain	Breakthrough logic: the project is completely geared toward the release of a new product

The different business units seek to expand their activities with an "always more" logic [DUB 93]	The projects seek to reduce the number of tasks that remain to be done [DUB 93], for the successful completion (end) of the project
Emphasis on individual performance	Focus on the collective performance, because the projects are designed to manage the interfaces

Table 2.1. *Working in professions and projects (adapted from Zannad [ZAN 08])*

2.2.2. *A job which is more interactive and more dependent on the downstream*

The arrival of project organization is accompanied by a transformation of R&D work with, on the one hand, an intensification of interactions that are now made with very diverse actors and, on the other hand, a certain loss of autonomy regarding the work. The downstream entities and logics are breaking into R&D work by being more present and more prescriptive.

In an organization by project, the R&D professionals are also interacting with the other functions of the company. This is an important change, because they worked until then in a professional space open to the outside but kept to the sub-environment of science and technology. Now, researchers and, particularly, the R&D engineers (the project form as described above being most used in the development phases) are in contact with the process engineers, production engineers, marketing professionals, finance, procurement, etc. R&D professionals, from now on, must interact with individuals who have neither the same training nor the same language, operational rationality and logic. They must learn to communicate, listen, understand and adapt to skills and logics which are different from their own. As a result of the projects, the R&D professionals are thus more in contact with the downstream entities as well as with various managers (project leaders, heads of the internal customer entities of the project as well as their usual business hierarchy). In some projects, the R&D professionals can also find themselves in contact with customers and suppliers who are integrated into design teams (as is the case in the car industry).

For R&D professionals, this is not only reflected by being in contact and by a need for more communication regarding their work, but it is a profound transformation in the way of working, because now R&D is more autonomous with regard to the activities they carry out. R&D professionals are involved in projects controlled by a project leader, according to a specification given by a client and/or other business service (marketing, finance, senior management), which strongly directs the work to be carried out and the results to be achieved. It is a major change compared to isolation and autonomy, which have long characterized R&D work. It is the outside environment that guides project management. Thus, they no longer have the autonomy. Instead, the R&D professionals are told issues and which problem to work on, according to the time frame: project deadline, milestones to respect, timescale granted for each stage, etc. Thus, project organization is translated by a loss of autonomy for R&D in terms of defining its own activities as well as organizing their own conduct and time management. At present, R&D activity takes place in a forced framework, in which the requirements are greater and there are ongoing assessments issued by the project leader, by different project participants in the various business units of the company, by external partners involved in the projects, etc.

Organization by project produces another significant transformation in R&D professionals 'way of working. For a long time they only interacted when they could present interesting and significant results, following a job that they considered as completed within their scope of intervention, but now things have changed. They will have to learn to show the work that they have produced more often, by presenting intermediary, partial results, not necessarily accomplished or completely assured – at the beginning anyway – which is very unusual and can be awkward and unsettling for these professionals. But how can we ensure that the professionals are working efficiently toward a satisfactory completion of a project? Through these very frequent interactions, we can ensure that the design takes into consideration the expectations and constraints of the downstream participants who support the logics of industrialization, of the customer, of finance, of the management of the product life cycle, etc., without which innovation cannot exist or succeed. It is often by showing what one does that helps to influence certain choices made in the project, on the condition that the person manages to convince the decision-makers of the relevance of the proposed technical option.

Thus, we are breaking away from the way of working that was certainly comfortable for R&D, where they were not exposed to logics, constraints, intrusions and evaluations from other business units, but which led to the abandonment of a large number of R&D projects. Indeed, R&D could devote significant resources in terms of time and financial means by developing far off the product or process prototype, which, having been developed by R&D actors isolated from those downstream, never met the interests, concerns, expectations or acceptable economic equations for the latter. This implies that if the project form reduces the autonomy that the R&D professionals had received until then, some of them will find a strong motivation in the fact that it increases the chances for them to see their work be used in new products or processes effectively adopted by external or internal clients. It is a source of recognition and motivation. It is particularly important for some professionals to know that they have contributed to the design of a product that achieves strong sales, for example. Some R&D professionals also appreciate working more collaboratively with very diverse business units. It is an important source of knowledge and skill development, and for some, this opens new prospects in terms of mobility and careers. Job changes toward the upstream business units are now more frequent.

In certain projects, according to the efficiency principles of concurrent engineering, R&D engineers remain implicated in the project until industrialization has reached an advanced stage, for example. Accompanying the project toward the downstream phases and seeing it succeed (and therefore not just knowing that it has succeeded) are again very important in terms of recognition and training. It is in following "one's own" project that one understands the difficulties of fitting a part, for example, in a more complex system, or the industrialization of a product, which proceeds from circumstances and physical and economic constraints, etc., which are very different from those encountered in the laboratory or in the prototyping phases. The first reactions of the clients – internal or external – are also extremely enlightening. The knowledge acquired is then very important for giving the R&D engineers new ideas. It is also important for them to select and guide, in light of this better knowledge, the downstream issues and constraints in industrial, economic, commercial and marketing terms.

The agile method, which is currently very popular, is a result of the effectiveness logics of the concurrent engineering in the world of software design. In the scrum methodology, for example, design teams work on short sequences, "sprints", after which the completed work is submitted for future users, or in any case, for actors who represent them. The rapid feedback on the work allows the choices made to be recorded, or alternatively to redirect them very quickly thanks to customers expressing their wishes or signalling inadequate usage while testing the pieces of product prototypes (which is easier to do in the world of software than in other sectors). If the team is coached by a scrum master, then there is only a little intrusion from a hierarchical management. The design team is very autonomous in the organization of its work (division and coordination of work, working time, etc.) and self-regulatory in the organizations that have truly implemented an agile mode to their processes. However, the retreat of the traditional hierarchical framework does not mean the absence of evaluation or pressure, but currently it is the figure of the customer who is omnipresent and who calls for effectiveness and efficiency. However, evaluation is much more prevalent than before, as each "sprint "leads to a work evaluation, and judgment of the work done (or not) is ongoing within the team.

The agile mode is one of the new methods of design that puts the client at the center of the discussion as well as the customers' added value of a new product and service. The method called user-centric innovation introduces the figure of the customer (real or represented) at the center of R&D work, where for a long time the scientific and technical challenge took precedence over the application and features of use.

This reconciliation of R&D and the customer is also a result of the decentralization of R&D from the headquarters of the company toward business units which are focused on markets and customers and are concerned with performance in the short term. Furthermore, many companies have conducted a partial or complete decentralization of their R&D, due to the willingness of the business units to have more control over the R&D work in order to direct them toward possible areas of challenge for them and to improve performance[1]. In the first configuration (which could be found, for example, in Rhodia, today Solvay, or in Thales), R&D is shared between one or several central laboratories and R&D laboratories in the business entities. The R&D center works on more upstream subjects that can potentially be broken down into applications for the different markets and business segments of the group. Decentralized R&D is more applied and more specific, working exclusively on processes and offers from the business units that host and finance it. Here, we have a form of structural ambidexterity, where there is significant differentiation in the roles of R&D actors and entities.

1 Bearing in mind that in a centralized structure, the head office finances R&D through withdrawals from the business entities that then pay for it but without being able to guide and evaluate its work.

The total decentralization results in the complete disappearance of central laboratories. These changes in structure are followed by a willingness to bring together the R&D of internal and external customers, in order to help guide its work and to increase the advancement of the work carried out in R&D toward the stage of usage. This has enhanced the dual phenomenon of intensification of R&D links with the actors, the downstream business units and logic, and its increased dependence on them. The projects carried out in R&D are the result of more or less urgent demands coming from the production, marketing, technical and sales departments or directly from the customers. Here, it is more difficult to sustain a real research project, when it weighs directly on the financial results of business entities, which are challenged on their short-term profitability.

All of these changes are part of a huge reversal of the innovation process and its management principles (see Chapter 1), from a science push model, in which R&D (and especially the R) was very autonomous and upstream in the innovation process, to a market pull model, in which R&D responds to the needs expressed by the clients or marketing services.

2.2.3. *Managerialization, bureaucratization and remoteness of technical work*

R&D executives – researchers and engineers – are currently based farther away from the technical activity. If we can relate this to their investment in new roles arising from the great changes in R&D management methods as well as, for example, the growing number of projects or the greater control granted to R&D over funds, then it is already a long-running trend. The organization of R&D work results in a clearer division of work between the design (performed by the executives) and the execution (performed by the technicians). Indeed, very quickly after their recruitment, young R&D engineers move away from the activity of "implementing" – the experimental activity, for example – to delegating it to the technicians. The latter are often the only ones who have a genuine expertise in some areas such as the control of scientific tools of experimentation, measurement, calculation, etc. They share the experimental work with the internship students, PhD students and post-doctoral students, and they form with them a community invested in development and knowledge transfer on these aspects of R&D work. This clear-cut sharing of activities between managers and technicians is partly due to the managers' "aspiration" to execute project management activities and more managerial roles. It is also due to an enthusiasm of young executives for these roles, which will lead them, quickly after finishing their studies, to move away from the more technical part of R&D work.

Recent trends toward managerialization reinforce this distancing of the technical work. All the executives are more and more absorbed by management activities: upstream negotiation of projects and budgets, presentations to management and customers, reporting activities during and at the end of the projects[2], etc. Those who take team or project responsibilities see the share of these activities grow to the point that they quickly abandon any operational scientific and technical activity (and in contexts where the knowledge is evolving very quickly, they can quickly find themselves unable to regain roles as direct contributors to R&D activities). The project leaders and managers are thus in charge of various roles: budgetary management, schedule management, day-to-day managing of the team, human resources management (evaluation, feedback on training needs, detection and prevention of psycho-social risks, etc.), passing of quality policies, management of relations with external partners, etc. It must be stated that a substantial share of R&D professionals has a real desire for these activities. Moreover, at present, some Master's students or thesis students manifest before even entering the profession the desire to do project management and innovation management, "without going through the background research part". Professional development toward project leader and project management roles are quantitatively important and arrive early in the career path.

These transformations in R&D work lead us to wonder about their consequences in terms of the dynamics of individual and collective skills. Do young R&D engineers spend the necessary time to acquire basic knowledge, core business, which forms the basis of their ability and legitimacy in this very technical universe of R&D? Aren't they risking "burning their bridges" by playing project leader roles too quickly, without having attained a strong enough skill base? How, at the collective level, can

2 The overturning of corporate funding of R&D toward funding by business entities has generated new requirements: R&D must, much more than ever before, argue for its benefits for the business entity to finance such a program of research, for example, and then it must constantly justify its use of funds with regard to how they are spent and especially with regard to what they produce or even bring in.

R&D professionals develop strong skills when the time devoted to the science decreases and when turnover increases? What does this mean for a profession when significant numbers of its members, and in particular, the new entrants, aspire to do something else quickly?

If most of the R&D professionals are satisfied with these prospects of rapid evolution toward roles of transverse or hierarchical management, it raises questions for those researchers who particularly aspire to develop their skills while continuing to work on R&D projects. For them, evolution toward project leader or manager roles is a form of abandonment, even if it contributes objectively to an improvement of the situation of the individual in the business (power, compensation, etc.). Some companies then offered various forms of payment to the R&D professionals who do not want to progress to management roles, which is the more traditional pathway in a company (see Chapter 4, Box 4.5).

Having said this, and even if this is still marginal, some businesses are currently experiencing difficulty finding volunteers to assume management or project leader responsibilities. It is true that requirements and pressure have increased sharply in recent times. This is not specific to R&D: proximity managers, who have not been trained in management, find themselves facing injunctions over the ownership of tools and complex logic. This is combined with a fundamentally sensitive position between the teams they have to manage and for whose results they are accountable and the general management. Recent pressures that combine performance injunctions and budget cuts accentuate these difficulties.

Finally, bureaucratization is another factor. The latter comes with a desire for tighter control over R&D, its spending, activities and productions. For example, R&D professionals are required to fill out time sheets, describing precisely how they have used their working time, based on different activities and in different projects. This is part of an internal logic of R&D financed by various actors (corporate, business entities, public authorities, such as the ANR (*Agence nationale de la recherche* – national research agency) in France or the European Union programs, etc.), each requesting to be able to verify that the volume of man-days that it has paid corresponds to the time actually spent by the R&D professionals on a project. Management control is much more present today in R&D, and it is about knowing which variable or fixed costs one should assign to a particular project, to calculate the sale price and profitability especially and/or to drive the design process to stay in the conditions initially planned. Bureaucratization is also currently linked to a very strong pervasiveness of quality standards and certifications, such as ISO, which imposes a formalization of the process, the establishment of procedures, with very detailed documentation to ensure traceability of the activities, the training received by staff, etc. In some sectors, we understand the rationale that there are very heavy demands: pharmacy, medical devices, aeronautics, nuclear, etc. The concerns around intellectual property, in a very competitive landscape, also lead some companies to ask their researchers and engineers to keep a written record of their work, in order to be able, if applicable (if the work leads to an invention), to provide evidence of the prior art of inventions necessary to obtain protection by means of a patent.

This bureaucratic trend raises questions about R&D innovation, which is meant to be managed through fluid structures based on mutual adjustment and a quick decision-making capacity.

Bureaucratization, as well as the increasing demands for justification of costs incurred *ex ante* and *ex post* evaluations, leads to paradoxical injunctions for R&D professionals. They should bring break-through innovations, while being able to describe what they would achieve and furthermore, to calculate the return on investment (ROI) of innovation to come, in order to obtain the necessary resources to begin to work; they must then create new knowledge, in the framework of a detailed schedule with specific milestones to meet; they must push for a renewal of the company offer, but must keep an eye out so that new technologies do not risk the system, which must be based on proven technologies and they must design in agile mode, while respecting the company's decision-making procedures, which provide multiple levels of validation and by continually reporting to their managers in structures that are from a hierarchical culture. One could imagine the difficulty in which the professionals are placed, since regardless of the choice they make, deciding one way or another, they will always be in an awkward situation because of the irreconcilable nature of these injunctions.

In the vein of these latest remarks, we begin section 2.3 by questioning the consequences of certain developments in R&D.

2.3. Current tensions and open questions as to the future of work in R&D

Some of the developments in R&D work which have been produced in the last 25 years have been well "absorbed" and can be a source of satisfaction for R&D professionals, at least for some of them (bearing in mind the variety of profiles and professional aspirations within this group). However, other developments generate tensions and disrupt R&D activities (section 2.3.1). Sections 2.3.2 and 2.3.3 focus, for their part, on more recent and still ongoing developments which profoundly transform R&D work and raise some questions, or even concerns, as to the future of R&D work.

2.3.1. *Increasing pressure and strong focus in the short term: how sustainable is this in individual and collective terms?*

For several years, R&D has both been put under unprecedented pressure and undergone a significant loss of autonomy, within a short period of time, because of the many factors previously outlined: organization by project; constant evaluation; strong time competition and obsession with the time to market; stronger dependence on business entities; conditioning of resources allocated to R&D, in particular based on its ability to demonstrate their usefulness *ex ante* and their performance *ex post*; focus on the short term, etc. It is not about making a case against these developments, some of which have produced important results such as a reduction in design deadlines and an increase in the transition rate of projects from the R to the D and toward industrialization and marketing. However, they raise questions relating to the sustainability (from an individual and collective point of view) of the new model, of R&D work (although we should state that conditions are different according to the companies, sectors, the R or D, etc.).

At an individual level, there are greater stress and broader and more different psycho-social risks (PSR), having already led to tragic consequences such as the suicides of R&D engineers. This issue of the PSR cannot be ignored at present, but it is the subject of very little academic work. Zannad [ZAN 08] has already pointed out that, on the subject of project organization, the literature is almost entirely focused on the beneficial effects of projects and, on occasions, it focuses on individuals and project managers. Although projects are a proven source of stress [ASQ 07], this organization is designed to focus attention on achieving project objectives, within specified time frames, that become increasingly short. The projects are great "demand machines", always asking more from individuals. The evaluation is constant, and it is the result of a multitude of participants: peers, members of the project team from other departments, project managers, and external partners, in addition to the business unit manager. While project management encourages, through physical proximity, communication and mutual adjustment, it also creates a lot of pressure as everyone is constantly under the gaze of colleagues, project managers and customers.

> Project management organization creates or generates major stress levels during the project (high pressure, ongoing and multi-faceted monitoring) and at the end of it (anxiety related to the lack of visibility as to what happens post development, disappointment or difficulties relating to the next assignment, etc.). During the project, stress exerts a very strong psychological pressure on the individuals who have the responsibility for the success or failure of the project. For the project managers, it is difficult to find a balance between project and personal life, especially because they are subjected themselves to an extensive schedule and struggle to stop thinking about the project outside of their working hours.
>
> The inspirational force, the knock-on effect and the strong pressure exerted by the projects are only positive up to a certain point, since over-involvement can have much more serious consequences and a much higher price to pay than the initial enthusiasm felt. Engineers also reveal: "I can no longer participate in a project the way I did with the previous project […] we're burned out after a project, we can't wait to get back to our jobs."

Box 2.7. *Stress generated by projects: the example of Renault (source: Zannad [ZAN 08])*

The generalization of the project model leads individuals to be in a continuous rotation between projects, without any time for a project break or time to work for the business unit needs, impacting the individual's capacity to recover. This recovery involves, on the one hand, a variation in work patterns that would reduce the psycho-social risks and, on the other hand, dialectic exploration/exploitation. In the project mode, individuals work based on existing knowledge (exploitation). However, the time spent in the business units can be very useful to maintain and develop knowledge, but it still requires time and resources to carry out activities of this type (which are not "cost-effective" in the short term for an internal or external customer).

Being confined to projects where there is no strong ambition for the creation of new knowledge can be frustrating for some R&D professionals, because this is not in line with their professional aspirations. Beyond this, there is the question of how relevant to the company's strategy is this focus on product development and the actions leading to their application (and thus to their generation of turnover). While many companies argue for the importance of innovation and the ability, besides incremental innovations, to take a more substantial lead on its competitors thanks to break-through innovations, this use of available R&D resources destinated for short-term accomplishement of development projects is questionable to say the least.

Studies (notably [BEN 98, BEN 06, CHA 01]) have shown how product management organization could be problematic in terms of the capitalization of acquired knowledge. During the project, everyone is under pressure and no time is allocated, either during or at the end of the project, to carry out a knowledge codification. In addition, the way that subsequent assignments on other projects is done doesn't take into account a previous reflection about the continuity of the knowledge base, the learning process, or the conservation of an organizational memory. The authors (see Chapter 5) have highlighted how projects deplete the expertise of the professions/business units, at individual and collective levels, endangering the company's ability to maintain its key skill areas and furthermore its innovation potential in the medium to long term. This risk is magnified in the case of long-term projects and in "output "projects.

Moisdon and Weil [MOI 97] reveal how team members express fear of losing their technical skills during projects. Indeed, they generally use one single technology. There is not enough time for knowledge development and it is not possible to exchange knowledge with one's peers. In addition, greater attention is given to economic issues than to the technical aspects of the work. Since the project is distinguished by its urgency, we know how to deal with any difficulty that has arisen in a previous project, but not necessarily the underlying reasoning, which is often hasty and flawed. This lack of capitalization is at the root of the disappearance of real product knowledge (in this case, the car) in the field of design, which is all the more pronounced since the business units are undermined by the project logic and do not deal with the fundamental issues anymore. Moving individuals from one project to another can prevent the business units from capitalizing on knowledge acquired in the projects and prevent individuals from acquiring new knowledge in their field of specialization.

Zannad [ZAN 08] recounts the feedback from Renault engineers participating in projects:

– "My priority is the objectives of the projects on which I am working; the other side of the coin is that professional capitalization happens after because my priority is to get the prototypes out in good time."

– "There is a fear of losing skills when working on a project because there is a gap in the technological follow-up given that, in a project, we are fixed on one technology and we cannot exchange skills with others on the project. With projects, it leads to a depletion of the business units and skills, this is clear."

– "The risk when one goes into a project is that as soon as you are cut off from the profession, you are no longer in the loop in terms of skills. You must strike a balance between projects, which demands a lot, and also trade, to keep up-to-date."

Box 2.8. *Projects and depletion of expertise: the example of the car industry*

It should be noted that once these difficulties were identified, some organizations developed practices to overcome them. This is the case in the field of knowledge management with tools and practices promoting knowledge capitalization [SIM 08]. HRM has also evolved, with, for example, attention being given to new key skills for work in the project structure, as well as tolerance to ambiguity, which is going to be sought among candidates at the time of recruitment or at the appointment of the project managers. The same applies to the consideration of the contributions made in projects in the annual appraisal of individuals, which is explained in Chapter 4.

Lenfle [LEN 08] warns of the dangers of extending principles of concurrent engineering to all R&D projects. Although concurrent engineering is very effective for managing development projects, for which it has been invented, it is counter-productive to apply it in the context of exploration projects, which are of a fundamentally different nature. Regarding these exploratory projects, it is essential to diminish the management constraints (which seek to chart, control and reduce uncertainty), to invent project management practices adapted to these phases of creativity, and a generation of ideas that are sometimes conflicting, etc.

However, it is even sometimes the ability to take on exploratory projects, which is currently in question in some R&D organizations. This is particularly the case in arrangements (very frequent now), where R&D is decentralized within corporate entities. These entities push for a greater focus on projects that can bring tangible results in the short term. And even when long-term projects are launched, it is difficult for the engineers to drive them forward when their time is drained and split because of urgent matters with projects that need to be completed quickly.

> Saint-Gobain Recherche chose to conduct the activities of exploration and exploitation within the same entities. This heterogeneity of projects between long and short terms is pleasing for most engineers, who see it as a source of enrichment and development, which shows how this allows a better understanding of the needs of the customers. However, these activities result from very different stances, skills and timescales, and there is a strong risk of sacrificing the exploration activities for more short-term projects. This can be linked to the engineers 'choices of ease and comfort or to the pressure exerted by the operational heads of corporate entities.

Box 2.9. *When the urgent takes precedence over the important. The case of Saint-Gobain (source: Dhifallah* et al. *[DHI 08])*

These limits push some companies to go backward in a way to recreate entities dedicated to exploration, which are generally placed under the responsibility of the corporate section (e.g. Explocenter Orange, now closed, or the LOF – Lab of the Future – of Rhodia-Solvay). While work tensions are less strong in these entities, where researchers can devote themselves to imagining the products and technologies of the future, they fall back with the difficulties that cause the poor connection between the upstream and downstream phases of the innovation process. On the one hand, the projects that are conducted in the exploratory entities arise from researchers' proposals and not from requests expressed by the corporate business units. It is then difficult to move projects to the development stage when this requires a handing over – especially financially – to business units who were not previously interested in these projects, and which are not included in their strategic agenda. On the other hand, even in the case of requests dating back from corporate units, the distance – physical, but also cognitive – between the upstream and those of the downstream actors creates the risk of conflicting processes between the work that will be conducted by the upstream and the reality of the expectations and constraints of the participating influencers of the downstream. In fact, little of the work carried out by the exploratory entities are included and valued by the divisions. This accentuates the mistrust vis-à-vis upstream activities, which are not well received in companies which are more than ever focused on short-term profitability. This short-termism really questions the ability of these companies to support intensive yet critical innovation strategies.

2.3.2. *Relocation, internationalization, outsourcing and open innovation: what is the future of R&D work?*

At present, R&D work is going through many transformations, some of which are still widely in progress or even foreseen. One example is the movement of R&D activities and investments toward the countries of the south. While the large industrial enterprises in the north have for a long time created R&D laboratories in other countries of the north, movement toward the countries of the south is more recent (see Chapter 1) and it is not clear how far this movement will go. While, for a long time, innovation was promoted by the countries of the north as the way to keep qualified employment on national territory and non-qualified employment destined to be delocalized, it is striking to see companies close R&D centers in their countries of origin[3], to open again particularly in China.

An initial set of questions refer to where the work of R&D will be in 30 years and who will be doing it. Also, what volume of R&D employment will remain in the countries of the north, in France for example? A second set of questions focus on the nature of the work carried out by the laboratories according to their geographical situation. For a long time, the international division of labor in R&D was clear. The laboratories located in Western countries, as well as in nearby countries, were conducting the more upstream work (for Thales, for example, the French and English centers). Laboratories in other countries of the north (e.g. the United States for Rhodia) enabled integrating localized research and innovation networks, which co-produced knowledge locally [JAC 11], but also enabled recruiting talent and, finally, better perception of the specific features of local markets, to ensure a close presence to the customers. The laboratories that were set up in the countries of the south concentrated on adapting products designed by the center to the foreign market specifications and providing technical support service for the problems encountered by the industrial customers.

3 These closures (also the R&D center of Sanofi in Toulouse in 2016) are at the root of social unrest, which is relatively new in R&D, which had been protected from restructure for a long time. The possibility of being made redundant as well as conflict constitute new dimensions in the work relationship of R&D professionals.

This international division of labor in R&D is changing, when one considers the phenomena of reverse innovation (Chapter 1) and that it concerns circulating quickly and transposing innovations designed "off shore" [RIC 15]. In the same way, some countries of the south have constituted ecosystems that are extremely rich in research and innovation, in which Western companies want to insert. What will be the role of R&D laboratories in northern countries, compared to those located in southern countries, in 30 years? How will these different laboratories belonging to the same company work together? We are noticing already how R&D internationalization results in more remote collaborative work, with the constraints and difficulties that time zones, languages, national and technical cultures, etc. give rise to. It also results in geographical mobility, which can be short (the length of a project) or longer, involving international careers.

Another important trend is that of outsourcing R&D. Large industrial companies "refer "in fact a part of the R&D effort to the external environment. This is for a number of reasons: lower fixed costs, reduced risk taking, exit from activities in which it is impossible to calculate *ex ante* profitability, ROI, etc. A first set of practices is to pay service providers (companies like Altran Technologies, small businesses, public research laboratories or technological platforms) to do what was done previously in-house. This permits others to carry out the R&D projects (or activities such as presenting samples, etc.) that are not done in-house, to have access to human resources without having to employ them, etc. A second set of practices consists of monitoring the research and innovation ecosystem to detect what is interesting in it before considering the ways and means of capturing it (recruitment of researchers and engineers, purchasing of patents or licenses, start-ups or competitors).

An initial consequence of these outsourcing practices lies in the evolution of the roles of R&D professionals in large industrial enterprises, in any case of some of them. In fact, they must monitor the environment, evaluate the projects and patents from public laboratories or competitors, identify interesting potential partners, report to the company management on this subject, begin collaborations, control research contracts entrusted to external actors, assess their work, etc. In sectors other than R&D, sponsors interested in outsourcing and subcontracting have pointed out the loss of expertise, which stems from it and which carries risks for the company, since its internal actors are no longer able to challenge the quality of services purchased externally. This can constitute a point of vigilance with regard to decisions concerning R&D and its perimeter: what to retain in-house and what to outsource. We should not forget the work of Cohen and Levinthal [COH 90] on the absorption capacity, which really showed how it is impossible – in a field as technical as R&D – to take advantage of what is done in the environment, without having the activities and important skills in scientific and technical fields in question, in-house. We should also note that it takes time to build skills in R&D and that it is not easy to go back on a choice of the disinvestment of a scientific and technical field.

The second consequence of this movement of outsourcing is a displacement of R&D jobs from the large industrial enterprises to technical services companies and SMEs. The work is different. First, engineers are assigned to projects conducted for and in customer enterprises, thus regularly changing the work environment or even business activity sectors and area of expertise. In start-ups and very small or small to medium-sized businesses, the organization of work is both more flexible and informal, with less specialization. Employment is also more uncertain, with limited career prospects internally.

These stronger links with various entities, private and public, geographically close or distant, participate in what is today called open innovation [CHE 03, Chapter 1]. What has changed in recent times is not the open character of R&D (see section 2.1.2), it is the variety in the types of entities that the R&D of large companies have relations with and the terms and conditions involved in setting up these relationships. Large enterprises interact with industrial groups, but also more and more with start-ups, intellectual service providers in the area of development, design, etc.; they maintain ties that are narrower and more multi-faceted with clients (end client, industrial client, real client, clients that are thought up and analyzed by marketing) at different stages in the innovation process (ideation, prototype tests, evaluation of the final product) and with different objectives (inspiration, legitimation, validation, etc.). It is often individuals or external communities (e.g. in the software or gaming world) who, not satisfied with just offering ideas or expressing their opinion on product prototypes, participate in the design of the products. R&D engineers have very different roles in such an environment. They set up mechanisms allowing customers to be involved, like design "tool kits "supplied to "lead users "[VON 01] or places dedicated to user testing. They make prototypes and allow testing and then collect and analyze the users ' feedback. This raises questions about the future of R&D internal workers, R&D professionals from their "historic" mission of contributing to R&D projects and these new roles open to the environment. Are they destined to become leaders of the lead users' communities or evaluators of the ideas generated by others?

Another transformation is related to the fact that entities and partners are scattered across the globe. Digital devices enable, for example, working collaboratively yet remotely, or allow one to collect ideas for product improvement from individuals around the world through the Internet via crowd sourcing platforms.

Recent times have seen an evolution from the practices of bilateral collaborations (see section 2.1.2) to multi-lateral relations involving many varied entities. For large industrial enterprises, it is about participating in the research and innovation ecosystems (localized or virtual, spontaneous or incited by the public authorities, etc.). They hope to take various resources from them: ideas, knowledge, access to technologies and equipment, recruitment pools, legitimacy, etc., in addition to a pooling of the costs and risks with regard to financing equipment and major R&D programs. In such an ecosystem, the R&D processes are now conducted across many organizational boundaries, in stakeholder chains or networks, and the interface work, interaction and coordination are now becoming a significant part of R&D work.

2.3.3. *The digital revolution: what is the impact on work in R&D?*

First, we must state that R&D has always maintained strong links with digital technologies, compared to other professions or business sectors. Not only digital technologies are designed fairly widely by R&D, but sometimes they have been designed first for R&D (including, but not only, in the military field) before being extended to the wider public. This applies to computers, supercomputers, the Internet, CAD tools (computer-assisted design), etc. Public debate currently focuses on the fears of future cuts in duties, professions and jobs due to the digital revolution, in particular, artificial intelligence, the world of R&D – which is equally affected by these threats – has a more "natural" and more grounded relationship with these technologies, which are also considered to be empowering.

In fact, R&D has been experimenting for years already with the incredible effects of various categories of digital tools (which must be classified according to their functionality). This is true in tools that allow a very fast and inexpensive flow of information: email, the Internet, electronic databases, several online scientific journals, social scientific networks, etc. In three clicks, we now have access to a volume of knowledge produced very recently worldwide, which facilitates monitoring and reviewing literature. The digital tools that enable very heavy calculations have revolutionized the world of research and design research. They have unbelievably shortened the calculation and data processing time of experiments, for example. They also allow the simulation of complex systems and sometimes completely replace any actual testing (as in the field of nuclear weapons, aerospace, etc.). They transform sometimes certain ways of designing research: in chemistry or biology, where a targeted approach of molecules based on a consideration of their structural characteristics has long been dominant, today's digital methods enable a blind screening of huge libraries of molecules. This changes not only the approaches but also the skills of scientists involved in these activities. These are new, scientific, multidisciplinary (such as bio-informatics) domains, which emerge and which reveal the strong potential for scientific and innovation breakthroughs.

Digital tools increase the efficiency of R&D as well (by accelerating calculation time or by reducing the necessary resources for testing). They also improve the reliability of its results, because the experiments have been replicated on a larger number of tests, because the computer-aided design tools leave less room for error than hand-made designs, etc. But these tools also transform how they work and they are the development support for new skills, ideas and technologies.

Virtual online development communities of specialized social networks, wikis, Internet forums, etc. are new spaces for brainstorming and comparing ideas, inspiring and training. Although this may raise questions regarding intellectual property protection for enterprises, they are unbelievable tools for stimulating the dynamics, which are already quite prevalent in these universes of sharing and exchanging within epistemic communities.

Digital technologies transform the working method within R&D teams and R&D projects. Collaborative tools facilitate collaboration, remotely or nearby. In R&D networks that are fragmented at the international level, in multi-actor R&D partnerships, these tools are invaluable.

Digital technologies modify and make work evolve due to the new opportunities that they open in terms of prototyping (fast and low cost). Prototyping is easier in the world of software (even though the software applications are becoming more influential in a large number of products) and tools such as 3D printers, which allow embodying ideas at an early stage and very easily. We can experience the user-friendly nature of an object, check its compatibility with another, etc., and other tools such as virtual reality and augmented reality also allow us to test user experiences at an early stage with regard to a future product. In this way, digital tools allow us to see much more quickly, more often and cheaply the intermediate states of production. This makes the project work easier in the agile method, as it allows very quick feedback from other business units and customers. Digital technologies offer immense possibilities regarding ways to engage clients (internal and external) in the design processes: crowd sourcing platforms, data collection on the remote use and exploitation of these, testing of physical or virtual prototypes, etc.

Certain industries that develop complex products (aerospace, automotive, defense, etc.) currently rely on very sophisticated integrative software platforms, extending the changes induced by the use of CAD tools and digital models. These PLM tools (product lifecycle management) contain databases and tools for collaborative work, which incorporate all the business units concerned. They can also be interfaced with external partners, including suppliers in sectors where the lead company (Renault, Airbus Helicopters, Naval Group, etc.) sub-contracts the design and manufacture of a very large number of parts, which must then fit perfectly into the overall system. These PLM tools facilitate the organizational and geographic breakdown of value chains and they can provide a valuable aid in terms of knowledge capitalization. They also lead to a greater formalization, with new requirements for information regarding the tool, for example, and one wonders if they are a support or an obstacle to creativity and innovation [PAR 18]. In any case it is certain that these tools change and will continue to change (although it is likely that they are changing and that they have gone beyond the first industries that adopted them) the border between those duties carried out by man and those carried out by "machine", leading to developments regarding the skills required in R&D and regarding the identity of these professionals.

At present, digital R&D is an overwhelming source of issues, which is placed high in the rankings of the company directors 'concerns (scientific directors and information systems directors) and a source of questions (if not worry) for R&D professionals. Digitalization is certainly going to change R&D work and the professionalism of researchers and engineers. However, uncertainty surrounds the depth and nature of these changes. For R&D managers, this calls for vigilance and considering how to go along with these changes in their work while encouraging the necessary training processes. The digital transformation invites researchers to push on with the initial work on these issues [BEN 16, PAR 18].

2.4. Conclusion

R&D work has changed considerably during the last 25 years, and profound transformations are still in progress, when we consider in particular the increase in the power of digital tools in R&D work. If the work is changing in keeping with processes in the scientific fields themselves, intensified growth strategies and innovation acceleration are powerful driving forces redefining research and development work. The advent of the project form, the pressure on the design deadlines, design-oriented use, open innovation and issues of ambidexterity and internationalization of R&D change the work context and the work content for R&D professionals. All aspects of the work are evolving: its objectives, the nature of its product, its timing, its spatial inscription, the skills that it requires, the actors that it engages, how value has been added to the final work, etc. These changes have an impact on individuals, professional identities, career pathways, etc. Therefore, recent advancements create instability and pressure on the R&D professions and professionals. Beyond these general trends, differences arise depending on whether we are talking about research or development, depending on the business sectors and the scientific and technical domains and according to the strategies and choices of enterprises regarding their methods for managing innovation and R&D. The following chapters will go into these factors and illustrations of R&D transformation in greater depth as well as the managerial responses implemented.

3

Rationalization and Creativity: R&D under Pressure

As Chapter 2 focused on the work of R&D and the changes that it has undergone, this chapter focuses on one of the main factors of work transformation in R&D: rationalization. While many R&D organizations are the subject of various types of rationalization operations, the aim of this chapter is to characterize these operations and point out effects that are sometimes counterproductive.

Rationalization practices have indeed become common in recent times, including in the world of R&D, which had long been preserved. If these practices have advantages that justify their implementation, then their impact on innovation, creativity and the performance of R&D teams, and therefore of the company, is a controversial and little-known topic. In this chapter, we first (section 3.1) conceptualize and present how rationalization has gained ground in R&D, thus leading to the reduction of different types of resources (human, financial, spatial and temporal), raising tensions and questions about their potential adverse effects on the innovation and creativity of R&D teams. In section 3.2, we focus on the understanding of creativity as a social process and the different factors that influence it, with particular emphasis on the importance of resource availability on creativity. In section 3.3, we examine the effects of slack reduction on the creativity of R&D teams. Finally, in section 3.4, we highlight the three mechanisms that explain the underlying effects of rationalization on creativity: focus of attention, ability to project oneself over time and leadership. In R&D organizations, it is necessary to take into account not only these risks but also the (potential) effects and rationalization mechanisms on innovation and creativity. This question is of the essence, considering the fact that it is hardly explored at present, and we will focus on this in this chapter.

3.1. Permanent rationalizations and reduction of available resources in R&D

Beyond the representations, it would seem that rationalization leads to adverse effects, which prove to be counterproductive compared to its ambition. These are complex processes that generate deep underlying tensions and hidden costs that arise over the medium and long terms. The purpose of this section is to first clarify the concept of rationalizing in light of R&D issues. Then, we define rationalizations in R&D as slack reduction strategies.

3.1.1. *The rationalization concept*

At present, new forms of slack reduction are taking place within companies, and it is important to understand them well while the very contours of this notion are unclear. In fact, the verb "rationalize" refers to the following similar terms: reduce, optimize and restructure, which do not carry exactly the same meanings. Thus, the term "rationalizing" is not clearly defined and its connotations vary according to the country and actors interviewed.

In France, for example, it is often associated with restructuring and layoffs and therefore with a very heavy significance, particularly from the viewpoint of employees and trade unions. In Anglo-Saxon studies, in which rationalization falls under the term downsizing, the relationship with workforce reduction is also present. However, downsizing has a wider sense. It can be defined as a set of activities undertaken by an organization's management to improve its efficiency, productivity or competitiveness. It thus affects the work force as well as the costs and work organization [CAM 94].

Despite these clear goals, many Anglo-Saxon studies show that downsizing has no significant effect on the company's economic and financial indicators, particularly regarding its profitability and productivity [BEA 12]. It would thus seem that downsizing leads to adverse effects that prove to be counterproductive compared to its ambition. These are complex transactions that generate deep underlying tensions and hidden costs that arise over the medium term, particularly in relation to the psychological contract and the loss of trust, motivation and loyalty of the remaining employees vis-à-vis their employer.

3.1.2. *R&D struggling with permanent rationalization*

Rationalization has been significant since the 1970s, but it has changed recently in terms of its targets, motives and justifications. From the 1970s to the 1990s, it targeted industry and production line workers, and then, progressively, restructuring has become "a permanent tool for companies seeking competitive advantage" [BEA 12]. At present, organizations are conducting "offensive" restructuring and not just "defensive" or "crisis". If the reasons for rationalizing vary, then it is clear that it has become common practice in both private and public organizations. It participates in the managerial toolbox mobilized to increase the efficiency of processes through the reduction of production costs, labor, time, etc. Rationalizing is thus about reducing the resources mobilized in these processes to improve their performance.

It is about creating light and flexible organizations that refocus on their core business for the benefit of greater agility. Industrial practices like lean management are also an illustration of the rationalization logic. This new managerial doxa is part of a society affected as a whole by modernity and social acceleration [ROS 03].

In R&D, lean management aims at obtaining a short time to market (time to market new products) and high overall costs efficiencies while minimizing waste. The new motto is to increase efficiency and accelerate innovation processes.

With the 2008 financial crisis, the sales of industrial companies along with investments in R&D dropped drastically (see chart 3.1). This crisis reinforced an already difficult context, in which R&D is strongly questioned and hustled, between acceleration of technological developments and intensification of competitive pressures.

Recently, we have observed how market and managerial logics have taken precedence over scientific logic [BOB 15]. Focus is clearly on the economic contribution of R&D, the acceleration of the R&D process to meet deadlines for marketing products and shareholders' interests.

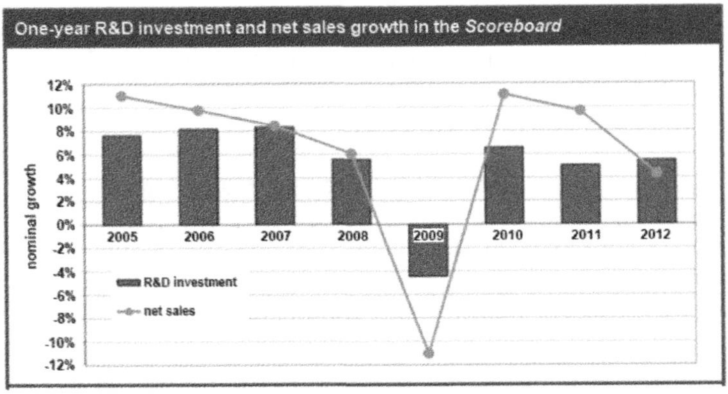

Figure 3.1. *Dynamics of investment in R&D, sales and profitability in EU countries*

R&D seems to be "taken" by many rationalization processes while reducing the consumption of resources, particularly time, whereas acceleration challenges are stronger than ever before.

These R&D rationalization processes are a source of tension, while knowledge workers are now subjected to a treatment that usually affects less-skilled categories. Even though R&D professionals are technological change drivers, they constitute a creative category that henceforth needs to be "stimulated" in order to produce best results and maximum profitability. They are closer to the archetype of the professional artist [MEN 03] trapped in an uncertain economy and exposed to the risk of inter-individual competition and new professional trajectory insecurities.

3.1.3. *Rationalization as a slack reduction strategy*

Slack reduction proceeds from a justification that companies are "fat", which means they have resources beyond what is strictly necessary to carry out their activities. In R&D, this leads to the conclusion that products can be designed with fewer resources than that consumed (and therefore wasted) by R&D currently. This notion of slack refers to unused capacity, overstaffing, excess budget and available time.

As explained by Beaujolin and Schmidt [BEA 12], the analogy with slimming diets makes it possible to better understand the idea of organizational slack. If it is in the zeitgeist to seek to eliminate these fats, which are considered excessive, the company that reduces its levels of "slack" recurrently runs the risk of anorexia and adverse effects that go along with it, especially in terms of weakening its socio-cognitive abilities to reflect, learn and act. Conversely, obesity also carries risks, especially in the longer term, and legitimizes strategic thinking on the corrective measures to be taken.

In this section, we will first explain the different dimensions of rationalization and slack reduction. Then, we will highlight the different tensions and questions raised by these different types of rationalization.

3.1.3.1. *Rationalization and slack reduction in different dimensions*

In contrast to the common representation of rationalizing or downsizing as being limited to staff or budget reduction (slack as human and financial, respectively), investments, space and time are also sometimes targeted [BOB 14]. From a longitudinal study of two cases, one in the music industry and another in the pharmaceutical industry, Bobadilla [BOB 14] shows that downsizing aims, directly or indirectly, at reducing slack in its different dimensions: human, financial, temporal and spatial. In R&D, this is expressed by a reduction in team size, budget, physical workspaces and time available for creation (see Table 3.1).

Slack dimension	Description of slack reduction operations
Human	Decrease in staff, layoffs, downsizing of teams
Financial	Reduction in R&D budgets Multiplication of evaluation criteria for R&D projects Reinforcement in the monitoring of budgets and expenditures
Time	Reduction in R&D cycles Tighter deadlines Pressure on results Acceleration of work pace
Space	Reduction of collective workspace Reduction of space for individual work

Table 3.1. *Types of slack reduction*

DrugLab is a research center, established in 1987, owned by one of the largest pharmaceutical groups in the world. In 2008–2009, all of the group's global research was reorganized into therapeutic units named discovery performance units (DPUs), with each DPU operating as a research center working in the first phase of pharmaceutical research. Research is organized with a group of people in the same location, with a small group (between 50 and 70 people) focused on a specific area. DrugLab, specialized in cardiovascular disease, is part of four DPUs: three in the United States and one in France. When the DPUs were set up, the site was identified as too big and, after restructuring, it changed from 100 to 70 people. Parallel to the social plan, in the same year, the old building was demolished and all the people of the research center moved to another building built nearby with open spaces. The teams started working with a 3-year business plan. Specifically, the smaller teams also changed the financial environment and moved into a new physical space. In 2010, there was another reorganization that immediately led to the departure of two leaders in biology and chemistry. After 3 years, the site does not deliver the results indicated in the business plan.

In 2011–2012, a strategic reorientation completely changed the working approach. Instead of closing the site, it was decided to convert it to a flexible performance unit (FPU). Previously, the center was in charge (from A to Z) of the choice of a possibility or target to enable the creation of a drug (for a disease, we must find the means to attack it) and R&D activities aimed at finding an effective molecule to treat this disease. Henceforth, the site becomes a provider within the group, it puts its business skills at the service of "internal customers" and it has to search for sufficient customers and activities to fund itself.

Box 3.1. *DrugLab: slack reduction in its different dimensions (source: Bobadilla [BOB 14])*

3.1.3.2. *Slack reduction as a source of tension and questions*

This R&D rationalization process causes tension and raises questions. Thus, while downsizing strategies aim at reducing the resources available for R&D, isn't it specifically the existence of a slack (and therefore of resources not totally absorbed by daily activities) that facilitates the creation of a team? Are these strategies, by breaking the inter-individual networks, not at risk of generating loss of key knowledge? Do they not jeopardize the knowledge assets that the company has accumulated over time? In the science and technology sectors that are the focus of this book, a company's competitive advantage is principally based on the resources and knowledge it possesses as well as its creative potential. Thus, slack reduction operations question their strategic relevance.

There are also natural limits to acceleration; we cannot notably increase the speed of cognitive processes of perception and processing of information by the human brain to infinity.

Ultimately, can R&D organizations indefinitely focus on reducing team slack without curbing innovation? Practitioners themselves are asking questions (see the testimony below).

> There is already a more global social movement that wants its desires fulfilled immediately and faster. Naturally, we agree that we do not understand why the research unit is so slow, or more exactly why is it not possible to delimit it using metrics. There is a need for acceleration... Acceleration is a basic postulate, even if it is only useful for the operational part of the research unit. Under this ambidexterity, is this acceleration also profitable for upstream processes and more research-oriented than development? More importantly, is it profitable?
>
> I do not know if slack reductions are the cause or consequence of this acceleration. It is certain that it is justified by the idea of a more present innovation on the market, competitors who are moving faster and faster and supervision as for the other operational entities of the company. Like any change, it creates fear and mistrust at the level of the team members, which lead to questions for which there must be answers now and immediately.
>
> More and more we perceive the application of all of the "lean" tools that we sometimes see in production and that we believe are useful in R&D. It always reassures people to have tools that work on other operational teams. It also facilitates the establishment of comparison bases and thus applies a global metric to the company in order to improve its performance. Specifically, how does this translate into practice? At present, we can no longer imagine a research unit that works without indicators. On a regular basis, we need to be able to know how long it takes to draft a plan and how much has been done "right from the first time", and we are challenged on these metrics. This enables us to increase our costs and accelerate our development time. The concern is that it is difficult for the teams to locate themselves in these acceleration processes. Even if putting under pressure is interesting and necessary, there are also high risks of inhibition for creativity, wherefore the measurement of efficiency and effectiveness is a challenge shared by many research unit managers."

Box 3.2. *Acceleration challenges (source: interview with Vincent Langlois, Manager of Poclain Hydraulics research unit)*

3.1.3.3. *Relationship between slack and innovation*

Cyert and March [CYE 63] were among the first to focus on the relationships between slack and innovation. They introduced the idea "success nourishes slack", thus eliminating the problem of scarcity and generating resources available to fund innovation. However, they also noted that, in times of crisis, slack is reduced and deployed to absorb the shock and be able to cope with a decline in profits.

Table 3.1 summarizes the relationships between slack and innovation as highlighted by several theoretical trends.

Behavioral theory [CYE 63]	Slack eliminates the problem of scarcity and creates resources to fund innovation Slack enables risk taking
Theory of the firm [WIL 81, JEN 76]	Slack accumulation is inefficient; it induces opportunistic behaviors Slack leads to funding unprofitable projects
Theory of resource constraints [BAK 05, GEO 05]	Remarkable innovations are possible with fewer resources

Table 3.2. *Relationships between slack and innovation: theoretical perspectives*

These theories convey two opposing views on the relationship between slack and innovation. It is emphasized that slack plays a crucial role in innovation by positively influencing behavior (innovation, risk-taking) and financial performance. The other viewpoint argues that there is a negative relationship between slack and innovation based on the fact that the existence of slack embodies and leads to a form of inefficiency. The authors of the theory of the firm, proponents of this second viewpoint, indicate how managers use slack to seek their own interests, for example, by engaging in unprofitable diversification projects. Similarly, resource constraint researchers make necessity the mother of innovation, while remarkable innovations are achieved especially in situations where resource constraints are high.

Bobadilla [BOB 14] shows that the same contradictory results are found in the studies related to the effects of downsizing on innovation. Thus, some show that the latter has no effect on innovation, while others assert the opposite. For example, Amabile and Conti [AMA 99] have shown that downsizing reduces creativity, defined as the production of new and useful ideas. They attribute this reduction in creativity to the deterioration of the work environment. Bommer and Jalajas [BOM 99] demonstrated that downsizing has a negative impact on the intrinsic motivation of R&D teams, but on the contrary, fear can mobilize individuals. In another study, Mellanhi and Willkinson [MEL 09] investigated the relationship between the effects of human slack reductions and the results of innovation (evaluated by patent). Through an analysis based on a sample of R&D companies in the United Kingdom and data collected between 1997 and 2003, they explained that downsizing levels are the key to understanding the effects. They also observe short-term negative effects that fade over time.

These contradictory results can be explained by the methodological limitations of these studies. First, slack reductions are often measured by considering only human and financial resources, or they can target other types of resources, including time and space. In addition, longitudinal studies are very rare, whereas the potential effects of this restructuring on innovation and creativity do not manifest themselves and can only be assessed over the medium to long term. Finally, the R&D-specific context has only been considered by a few authors, which makes it difficult to extend to this particular professional universe results obtained from other perimeters.

The ability to be creative is of interest to scientists from different disciplines, practitioners and citizens in general. However, there is no simple definition of creativity that encompasses the different dimensions of this phenomenon (individual attribute, process through which new ideas are generated, social construction). In the following section, we shall try to clarify this notion in light of R&D teams.

3.2. Creativity: between individual attribute and social process

In the first section of this chapter, we have discussed the complexity of rationalization and considered some negative effects of slack reduction on the organization and individuals. We have underscored the lack of sound knowledge relating to the effects of downsizing on creativity and innovation in the R&D-specific context. We have revealed other forms of slack reduction that are not limited to reducing conventional resources (human and financial). Furthermore, we have shown the advantage of understanding rationalization as slack reduction strategies acting on different dimensions (human, financial, temporal and spatial). We have seen that "slack" is a key element to foster creativity and innovation. Access to a certain level of resources is a necessary condition to encourage the creative behavior of employees [AMA 89]. In the following sections, a review of the different creativity dimensions is presented. First, we will explore individual creativity (section 3.2.1) and then show the importance of understanding creativity as a social process in R&D contexts (section 3.2.2).

3.2.1. *Individual creativity*

3.2.1.1. *Cognitive science perspective*

Academic research on cognitive science suggests that our thinking is not spontaneously creative; it rather has a natural tendency to stay in the most usual ways. It also suggests that creative individuals are able to connect between ideas that less creative individuals are not able to connect [POI 13, WAL 26].

For many years, cognitive science studies have focused on understanding brain activity and behavior.

Initially, creativity is known to be associated with the right hemisphere of the brain [ROC 94]. However, other researchers have indicated the relationships between creativity and the left cerebral hemisphere [MIH 10]. Over time, studies have shown a similar duality in the process of creative thinking, depending on a combination of irrational and rational thinking or "divergent" and "convergent" thinking processes. De Bono [DEB 82] distinguished "lateral" and "vertical" thinking.

More recently, neuroscientists Fugelsang and Dunbar [FUG 05] have noticed that people pay more attention to evidence regarding plausible theories than implausible issues. This means that our first intuitions would be based on the most common solutions in order to enable us to respond quickly to situations encountered on a daily basis. According to Edward de Bono, training our minds to think outside the box and take new directions is lateral thinking, which therefore requires effort and learning. Thus, they discovered that when the data are coherent with the theory, the neural mechanisms in a region surrounding the hippocampus as well as the learning and memory center of the brain are activated. When data are incoherent with a person's theory, brain mechanisms in the prefrontal cortex involved in error detection, attention and monitoring of conflicting viewpoints are activated.

While neuroscientists have chosen to emphasize the activity of the brain during creativity, clinical perspectives emphasize dialectical elements expressed as "rational" or "irrational". Poincaré's "conscious" or "subconscious" model of thinking explains that the tension between these types of thinking generates a new idea, not a dependence of one on the other. This implies that the "creative" individual can operate with both polarities, for example, through enjoyment and discipline or passion and objectivity. Mihaly Csikszentmihalyi [CSI 96] described the creative person as a sort of "man with all" who has great energy but who is often calm and at rest; someone who is intelligent but naive at the same time; someone with a combination of play and discipline and someone who is humble but also proud. For Freud [FRE 76], creativity was like an expression of unconscious. In the Freudian paradigm, creative behavior is considered as a more socially acceptable form of undesired subliminal unconscious conflicts or even neurosis. Other researchers like Wallace and Gruber [WAL 89] have considered creativity as more than a single, multi-causal and interactive individual disposition, evolving in relation to a range of opportunities and influences.

3.2.1.2. *Personality of creative people*

The five-factor personality model (known as the "Big Five" model) has been a reference model for more than a quarter century [DIG 90]. It can be represented by the acronym OCEAN (openness, conscientiousness, extroversion, agreeableness and neuroticism). These successively designate openness to experience in relation to family satisfaction, conscientiousness of indifference, extroversion in relation to introversion, agreeableness against hostility and neuroticism in relation to emotional stability. Of these five main personality traits, neuroticism should be specifically mentioned here, as Gotz and Gotz [GOT 79] noted that the correlation between neuroticism and creativity was negative in sciences, but positive in arts, and other researchers did not find a significant correlation.

On the basis of other studies, Amabile [AMA 83] described the following personality traits that characterize creative people: open-minded, independent judgment, self-reliant, confident, attracted by complexity, aesthetic-oriented and risk-taking. Creative individuals also tend to be discovery-oriented, which makes them apprehend situations from multiple perspectives, find problems and ask new questions.

In research on scientific and artistic creativity, Feist [FEI 06] gave a clear portrait of the creative personality of both groups (scientists and artists). He noted that creative people were more autonomous, introverted, open to new experiences, diligent, ambitious, dominant, hostile and impulsive. Among these characteristics, openness, awareness, self-acceptance, hostility and impulsivity seem to be the most important. The study showed that creative people in arts and sciences do not share the same personality traits. Artists are more distinguished by their emotional instability, coldness and rejection of group standards. Creative scientists presented low scores on socialization scales such as responsibility, socialization, good impression and conformism.

Thus, we see that creativity is related to a cognitive-perceptual style that includes the collection and application of diverse information as well as the heuristic processing of information, good memory and ability to remain focused for long periods of time [AMA 89]. In addition, traits such as research and problem formulation, combination and assessment of ideas also seem to be important.

Other elements are also important. Creative production depends on an extensive and advanced knowledge base in a given field. Experience in a field is an essential component of employee creativity because an individual needs a certain level of familiarity to do creative work. Furthermore, creativity also requires a certain level of passion or inner strength that drives individuals to persevere when faced with challenges that are inherent to creative work. Several studies on individual creativity have focused on the importance of intrinsic[1] motivation for creativity [AMA 89, SHA 91]. Furthermore, intrinsic motivation is critical for the creativity of R&D professionals [BOB 14].

1 It is worth recalling that there is intrinsic motivation, when the action is conducted solely by the interest and pleasure that the individual experiences, without waiting for an external reward.

Finally, the subjective experience of people at work is very important, like the emotional side of their organizations, not only because the positive affective experience is related to outcomes such as job satisfaction but also because the affect is directly related to how people think about their work site. There is a dynamic relationship between affect and creativity. Affect is not only an antecedent for creativity but also a direct consequence.

3.2.2. *Creativity as an idea production process*

In addition to the individual viewpoint, creativity can also be analyzed from the process perspective. Emphasis is no longer on the individual, but on the production stages of creative ideas. Thus, creativity can be understood as a cycle that starts with the generation of alternatives to solve a problem and ends with the exploration and evaluation of these alternatives in order to respond to the initial problem.

The analysis of creativity processes began with the following linear stages of Wallas [WAL 26]: preparation, incubation, illumination and verification. Amabile [AMA 88] identified five stages of problem solving: presentation, preparation, generation, validation and evaluation. According to Newel and Simon [NEW 72], problem-solving is a complex process that includes a problem construction phase, information encoding, selection of a problem-solving method and its application. As a result, Runco and Sakamoto [RUN 99] provided a two-level model in which primary processes interact with secondary processes. It is interesting that, in the course of ideation, all critical, rational and convergent thinking is deliberately deferred in favor of an irrelevant divergent and irrational thinking, where all options are possible. During evaluation, everything turned out in the opposite direction.

The above-described models tend to assume that a problem, task or objective that requires creativity already exists and that a creative process must be applied. However, practically, creative behavior can occur before a problem is available to be identified or formulated and continue until the action required to implement a solution is taken. In order to take this fact into consideration, Basadur *et al.* [BAS 13] proposed an improved four-stage model.

1) Generation: the first stage in this creative problem-solving process is proactive acquisition and generation of new information as well as detection of trends, opportunities and problems. This is what Simon [SIM 77] called "opportunistic surveillance". Here, the physical contact and involvement in real-world activities will alert the individual against inconsistencies and difficulties. These inconsistencies are later on used to suggest new problems, identify improvement and innovation opportunities as well as propose projects that are worth undertaking. In this stage, problems and opportunities are acknowledged but not yet clearly articulated or understood.

2) Conceptualization: in this second stage, a problem or opportunity identified in the previous stage is analyzed to create a complete conceptualization or model of the problem area. In this case, understanding of a problem is achieved not through direct experience but through abstract analysis. This conceptual knowledge is later on used as the basis of ideation by which one or more solutions for the problem are developed.

3) Optimization: in this stage, conceptualizations of the previous stage are criticized against the constraints of the real world to identify practical difficulties. Alternatives are systematically reviewed to develop a plan for implementing an optimal solution that can be carried out with existing resources.

4) Implementation: this stage completes the creation process. The cognitive activity in this stage was to experiment the new solution, evaluate the results and make adjustments if necessary for successful implementation.

For many years, cognitive approaches, personality-based approaches as well as process perspectives were the most common. The aforementioned studies suggest that a creative personality exists and creative thinking follows a process. Psychology and cognitive science have contributed to our understanding. Creativity seems to be an affect-filled event, in which complex cognitive processes are shaped and mixed with emotional experience, skills and knowledge.

A critical limitation of these approaches is that these visions ignore the social contexts involved in creativity. In the following section, we will study creativity as a social process and the environmental factors influencing it.

3.2.3. *Creativity as a social process*

3.2.3.1. *A social and interactive approach to understanding the creativity of R&D teams*

According to the social and interactive theories of creativity [AMA 96, CSI 96], a creative process is the result of not an individual, but rather the interaction of the individual with his/her social and organizational environment. In R&D environments, this process develops and unfolds in a social context. The productions of this process are ideas that do not only have value but are new and useful. Context characteristics are the dimensions of the work environment that influence an employee's creativity but are not part of the individual, for example, job content, relationships with leaders, other team members and available resources.

Scientists and R&D experts rarely work on their own. Organization by project is now customary in this context. By analyzing the dimensions of the teams' creative work, we observed that research, filtering, experimentation and cross-fertilization take place collectively. At the program levels, each "creative" project must be validated by gatekeepers or reference communities that will define the value of this contribution. According to Csikszentmihalyi [CSI 99], creativity must be understood in relation to other aspects of the environment in which the teams work: area and field. The field is a necessary component of creativity because it determines what is innovative compared to the old. Creative production mostly depends on the team and an advanced knowledge base in a given field. Experience in a field is an essential component of the teams' creativity, because they need a certain level of familiarity to do creative work. Thus, we observed the importance of understanding creativity as a social process and analyzing the different factors influencing it.

3.2.3.2. *Factors that influence creativity*

Contextual characteristics potentially influence the creativity of individuals and teams. Literature distinguishes three types of contexts, namely organizational, social and work-related. Amabile's work [AMA 88, AMA 96] work served as a general framework and described various relevant factors that influence creativity: (1) organizational encouragement and support to creativity (e.g. support new ideas, acknowledgment of creative work and encouragement to take risks); (2) encouragement of direct superiors (e.g. setting clear objectives and opening up to new ideas); (3) support work group (e.g. open communication and trust relationship within the group); (4) sufficient resources (such as facilities, money and information); (5) a job perceived as challenging and, finally, (6) the freedom to decide on how to do one's job. Amabile [AMA 96] showed that low-creativity projects are often linked to organizational obstacles (e.g. political problems, over-criticism of new ideas, destructive competition and emphasis on maintaining the status quo as well as emergency pressure).

Finally, the psychological climate defined as the set of individual perceptions of policies, practices and organizational procedures that influence team behavior is another critical factor that may favor or hinder creativity.

Thus, we realized that understanding work environment factors contributes to a complementary understanding of the creativity of R&D teams. Within this framework, it was crucial to examine how slack reductions interact with team creativity over time. Of the factors discussed above, the level of available resources seemed to be an important variable that deserved further analysis. In order to generate innovative products, organizations are deploying and increasingly organizing themselves in the form of project teams. However, these teams did not always manage to meet the expectations of creativity and innovation because they could not benefit from the resource slacks that they needed in order to be creative. Recently, R&D has undergone an intense transformation that has led to the slack reduction and standardization of processes. How do these slack reductions interact with the lives of teams and organizations?

In the next section, we study the effects and interactions of different types of slack reductions on team creativity.

3.3. Ingredients and negative effects of slack reductions on creativity

As we explained in the previous section, the creative behavior of a group is influenced by the available resources and slack levels, which we define as resources that are higher than those needed to produce the necessary results [BOU 81, CYE 63]. Slack levels are therefore factors for understanding the creativity of R&D teams, but this importance has often been underestimated in practice. In this section, we will first explain the process-based vision of slack reductions to show the different components at stake. Then, by mobilizing its components, we will show the effects of human (see section 3.3.1), financial (see section 3.3.2), temporal (see section 3.3.3) and spatial (see section 3.3.4) slack reductions to later explain the mechanisms that justify this relationship.

3.3.1. *Slack reduction components*

Slack reduction operations take place in the long run. They result from multiple and interdependent causalities that combine over time, and they produce major changes for R&D teams and interact with components of several types: slack reductions can be considered as a process. According to Mendez [MEN 10], the historical nature of a process means, on the one hand, that it is integrated into a context where certain elements or components (situations, events, actions, etc.) play an active and dynamic role in the process in a contingent and sustainable manner. The different forms of slack reduction themselves interact with five types of components: contextual, procedural, relational and emotional, cognitive and activity-related [BOB 14]:

1) contextual: including the economic context, industrial culture of the organization and history of the organization;

2) procedural: how to implement slack reductions (means of implementation and type of targeted resource);

3) cognitive: relating to the mental processes of perception of work environment and attention;

4) relational and emotional: social and emotional relationships within teams and their leaders;

5) activity: nature of research, control level in R&D processes.

3.3.2. *Human slack reduction effects*

Slack reduction can take different forms and is a powerful process that interacts with many types of components. The slack reductions focusing on "human slack" are crucial in their ability to trigger negative reactions and emotions, especially when they are poorly implemented.

These human slack reductions are called layoffs, restructuring, downsizing, etc. The French system of social regulation of restructuring, specifically represented by the PSE (plan for safeguarding employment), is both complex and complicated [BEA 12]. Layoff procedures are governed by the law or constantly changing laws, which specify the conditions for consultation as well as negotiation and introduce obligations for companies to reclassify. However, despite this legal coverage, layoffs are often poorly managed, which provokes various reactions from laid-off workers or survivors (people who were maintained in the organization).

3.3.2.1. *Poorly implemented slack reduction produces negative reactions*

The way in which human slack reductions are implemented (legitimacy of the decision and justification of job cuts, sense of procedural justice, communication throughout the procedure) moderates the reactions and behavior of the teams over time.

The human slack reduction processes are considered fair when they provide consistency in terms of the targeted individuals and time, are free from bias, integrate and reflect the views of those concerned and comply with moral and ethical standards. When human slack reductions are considered as fair and legitimate, team behavioral responses can be accelerated toward the trust that the organization can endure and new relationships can be created.

In the course of time, positive reactions to change can lead to the re-establishment of relationships between team and work members. For example, after several months of PSE, the levels of stress and bereavement can reduce and teams can return to work more serenely. On the contrary, if human slack reductions are considered unfair, then the layoff decision is neither justified nor legitimate, and if there are no structural changes in the activity (change of positions, levels of responsibilities, new missions, organizational structure of the organization), then negative reactions (anger, fear, stress, sadness) are produced that can lead to low levels of trust, high levels of conflict and disruption of human networks. These negative behaviors affect the work environment and are very harmful to the generation and sharing of ideas and intrinsic motivation.

The way in which human slack reductions take place, the types of resources targeted and the manner of implementation (timing, procedural justice, communication, leadership) are key to explaining team reactions. Within the framework of DrugLab, the timing of the first PSE was considered as long and the layoff rules and criteria were not respected. Also, in both phases where human slack reductions occurred, communication from management was considered as insufficient. There was little communication, and management took decisions secretly, without the involvement of team members.

The selection process of individuals and selection criteria of laid-off persons were considered unfair. At DrugLab, poor procedural justice has led to negative reactions (anger, fear, betrayal, low morale) and low self-esteem, particularly among "survivors" with high organizational commitment.

Illustration of reactions to human slack reductions:

– "Layoff rules were not clear from the start. They chose those they wanted to layoff, regardless of the social criteria that legislation may impose. They fired everyone so that you apply again for a lower position. Knowing that they were already aware of the person they were going to employ" (Researcher, chemistry).

– "For this reorganization, all positions like technicians were canceled, so it was necessary to apply, redo CVs and job interviews. It's weird, after 22 years in a company, going back through this process is difficult. They had recreated positions with very specific job definitions. Practically, there was a name per position. This is quite disturbing, especially for those who have been laid-off, even for those who remained in the company. We want to keep people and we want to layoff others" (Research Associate, Pharmacology).

– "This marked the beginning of a disruption in the very personal investment for the company. I think most people now come to work because they are interested but they will no longer have any emotional relationship with the entity" (Researcher, Biology).

– "It is clear that during these periods, motivation is nil. We lose confidence and any point of reference, we do not really know why we work (Researcher, pharmacology).

– "If we look at things very concretely, there was a business plan that had been written. Thus, you think this business plan has to be drafted but with 30% less resources. It's a bit of a panic. This means that there is both a diminishing motivation and at the same time a kind of fear of not being able to deliver what we have promised to do" (Researcher, chemistry).

Box 3.3. *Key procedural component at Druglab (source: Bobadilla [BOB 14])*

3.3.2.2. *Overly frequent human slack reductions have adverse effects*

Slack reductions affect relationships between individuals, team and leader. Most importantly, the dismissal of a central member (i.e. team leader) without any legitimate justification is a negative factor that affects the morale as well as the interaction and connectivity levels between the groups. Connectivity levels between members are changed because information exchange is less fluid due to the departure of some central members. Furthermore, Susskind [SUS 07] highlighted the difference between voluntary and involuntary changes in the network. As part of the involuntary changes, the network recovery process was longer. The underlying mechanism of this network deterioration is access to knowledge through relationships.

In our case, we observed that "human slack" reductions lead to the deterioration of the work environment. In fact, the way the dismissals were organized led to an increase in conflicts and mistrust among members, which reflected as retention of knowledge. All of these negative factors are detrimental to the generation and sharing of ideas and intrinsic motivation of individuals and teams.

In the context of frequent human slack reduction, R&D teams face difficulties adapting to changes, which can create behavioral rigidity (less flexible individuals).

Affect and positive emotions have a systematic impact on performance in many cognitive tasks and emotions; furthermore, a good work atmosphere encourages creativity. On the contrary, poor climate and negative emotions inhibit creativity [BOB 14].

3.3.2.3. *More stress and conflicts within R&D teams*

Some stressors, such as emergency or external competition, may increase creativity while other stressors, such as time pressure or intra-group competition, may reduce it. We noted that bad implementations of human slack reductions increase competition as well as intra- and inter-group conflicts and thus destroy the sharing of ideas. However, conflicts differ in their form, cause of discord and effects produced. We identified [BOB 14] the following three main categories of conflicts that occur during human slack reductions:

– task-based conflicts, which concerns discussions and debates on work in progress;

– relationship conflicts, which concern interpersonal interaction between group members;

– process-based conflicts, which relate to the way work is done (and assignment of roles and responsibilities).

These conflicts are related to restructuring for several reasons, which are very well described by Cameron [CAM 94] (see box below). Organizations, after "human slack" reductions, witness an increase in resistance to change, and the climate becomes more political. Furthermore, loss of confidence plays a vital role in increasing conflict because there is more mistrust among members. Relationship conflicts threaten the survival of the team. Distress and animosity among members encourage withdrawal and affect the sharing of ideas. Process conflicts affect the ability of the team to plan, visualize results and perform creative tasks since team members do not often have a clear vision on how to go about work or who does what. All of this limits the relationships between the different team members. Above all, the anxiety generated by conflicts affects the different stages of creation through concentration problems and individuals are focused on conflicts, which does not make the mind creative.

1. Centralization

2. Crisis mentality, short-term

3. Loss of innovation capacity

4. Resistance to change

5. Degraded climate

6. Strengthening specific and politicized interest groups

7. Non-prioritized choices

8. Declining trust

9. Increasing conflict

10. Restricted communication

11. Lack of teamwork

12. Lack of leadership

Box 3.4. *The 12 declining organizational problems according to Cameron [CAM 94]*

3.3.2.4. *Loss of skills and knowledge*

In slack reduction situations, in the case of DrugLab that we studied, scientists hid information and knowledge. As soon as the results obtained endangered the continuity of their research project, they preferred concealing the results. It is important to note that the retention of knowledge is different from the lack of sharing because, in addition to the omission of sharing, it incorporates the intention to refuse to give information that someone else has requested. This intentional retention of knowledge inhibits the creativity of colleagues and generates the same effects of mistrust and non-solidarity.

Thus, we can conclude that downsizing leads to the loss of knowledge and has a negative impact on the ability of companies to benefit from shared knowledge through the interpersonal relationships that are essential for the creative process. Certainly, the generation of ideas begins with an individual; however, as part of a team, the sharing and evaluation of ideas is influenced by information exchange. It is both an explicit and implicit process. An individual in a team can draw attention to an aspect that others have not thought of. This explicit process is done through an open communication of ideas. From the viewpoint of social and interactive creativity, disruptions caused by human slack reductions, in a scientific context, are particularly disadvantageous for the sharing and evaluation of ideas.

– "First try to save one's bacon, put oneself first before the others. This created distances amongst people" (Researcher, biology).

"We lose knowledge. I worked with people who are no longer part of this team so we had to rebuild teams. Certainly a waste of time. Sometimes the level of motivation to tackle issues, where there may be 1 or 2 people out of the 5 or 6 who constituted the team, is not very good. We lost 4 of our colleagues, we have 2 people left" (Technician, chemistry).

– "So there is loss of knowledge as well as loss of time. Sometimes all these end up being a demotivating factor. Those who initiated these projects are gone. We are not the experts, we were already helping these people. How to react better, how to bring in positive things. It's difficult because when you see the experts who have left a project, you start thinking that maybe the battle is already lost. I think we all went through this kind of phase where there was a lot of loss of knowledge, reorganizations of teams after global reorganizations, they were not the same people anymore. And later on rebuilding trust" (Researchers, DMPK).

– "The major change is still the PSE that has confused spirits and left some traces. This has created problems that still exist" (R&D Manager).

– "The effects are reflected by a lack of motivation. People have become embittered, they are less fun than before. There have at been a lot of changes" (Researcher, chemistry).

– "It's very funny, despite everything there is a lot of emotion. For a discovery center, we could say that everything is based on science and data, finally we realized that there are many things that are guided by emotion, especially here considering that people have been there for a long time and because they built this center. Everyone feels like having contributed to the project. Everyone values each other and values the center because they have been there for 15 years. They witnessed everything right from the beginning" Trust had to be rebuilt" (DMPK Researcher).

Box 3.5. *Illustrations of the effects of human slack reductions (source: Bobadilla [BOB 14])*

3.3.3. *"Financial slack" reduction effects*

3.3.3.1. *Slack reductions increase control levels*

Bobadilla [BOB 14], Nohria and Gulati [NOH 96], Love and Nohria [LOV 05] and Daniel *et al.* [DAN 04] highlighted that the relationship between the creativity of R&D teams and available financial resources (budgets, investments) is curvilinear. Creativity increases with the provision of financial resources, up to a point from which it begins to decline. In fact, when financial resources become (too) abundant, there are phenomena (lack of control in expenditure, laxity on the results produced, useless investments) that impede the generation of ideas and relevant evaluation of these ideas. Conversely, a significant and continuous reduction in R&D budgets and investments leads to increased requirements and control levels relating to the evaluation of R&D projects that inhibit risk-taking and generation of creative ideas.

3.3.3.2. *Frequency of financial slack reductions affects risk-taking*

Financial slack reductions can have negative effects on R&D creativity, even though we have already highlighted the univocal relationship between financial resources and creativity. Andrews and Farris [AND 72], for example, showed how financial resources and facilities are not essential factors in the performance of scientists. Very creative results can be achieved with limited financial resources [BAK 05], provided the emotions, climate and relationships between team members are good.

On the contrary, at a certain level of budget constraint, the effects are negative on creativity: teams have more difficulties coping with the continuous lack of financial resources and the stock of "cheap" ideas starts reducing. The lack of financial resources can also urge experts and talents to leave the organization in search of other companies capable of providing them with financial support to develop their projects under good conditions.

3.3.4. *Temporal slack reduction effects*

3.3.4.1. *Research activities on standby and decrease in intrinsic motivation*

Rationalization leads to slack reductions in its various dimensions that intermingle. Thus, human slack reduction operations are also reflected through work time effects and the way in which it is assigned to various activities. During human slack reduction periods, the research activity automatically stops. On the contrary, the time allocated to discuss with colleagues and reflect on the current situation highly increases, and questions are raised and debates ensue concerning the degradation of the work environment, conflicts, the degree of uncertainty as to who will be dismissed, etc. These preoccupations are time-consuming (thus inducing temporal slack reduction) and negatively affect cognitive processes and orientation (focus of attention) toward creation and creativity.

In these phases, individuals truly devote very little to science, resulting in a very strong decrease in intrinsic motivation that is normally fueled by the search for novelty and challenges, exploration, ongoing learning, etc.

3.3.4.2. *Excessive time pressures generate conflicts of attention and a change of focus*

When teams experience extreme time pressure, attention is focused on the most effective and efficient (rapid) way of carrying out an activity, to the detriment of creativity.

This observation is not new because, in 1972, Andrews and Farris demonstrated an inverted U-shaped relationship between scientists' creativity and time pressure. This means that high levels or, conversely, low time pressure, have a negative impact on creativity. Similarly, they pointed out that the most creative scientists were those who had to cope with above-average time pressures, in active communication with their colleagues, motivated by their work and involved in certain administrative tasks as well as technical activities.

In contemporary work contexts, R&D professionals participate in a growing variety of activities, tasks and roles simultaneously. This leads to "vertical loading" or "polychronicity", which is defined as the performance of two or more activities simultaneously [BLU 92, HAL 83]. Excessive short-term focus shifts to explicit knowledge rather than tacit knowledge, reducing the ability to be creative.

Also, the question is focused not only on the amount of time available but also on its quality. With slack reduction, R&D creative people and workers express the need to have creative windows, which allow a step back to open up their minds for creativity.

– "In biology, we were trying to understand certain things; they were put aside in order to focus on the more important things, which related to the main goal. We eliminated a little of what was not nice to have but what was improvement, we gave up. We focused a little bit more on the goal that was set" (Researcher, chemistry).

– "The timing is very different. The Flexible mission is completely different, the timings are very short. Science is flexible. It's a huge challenge. When we have projects coming from everywhere, we need to be an expert within three weeks. Adaptation must be total and immediate for someone who comes for inflammation and another for oncology. It's very different" (Researcher, biology).

– "This new way of working will probably have an impact on creativity in all its senses, it is more oriented with all the defects that go with it. It will be more difficult to create its place and appropriate things. And I think that if we do not take ownership of our research projects, it will be more complicated. We shall see, I do not want to be negative for the future, I think it's not obvious" (Researcher, chemistry).

– "Now it may be a little different because we have the impression that missions that will be shorter in time but much more focused will be requested. This may involve, for example, only biologists or chemists. What may be missing after this is the possibility of a decline in cohesion between the different departments" (Researcher, chemistry).

– "Timing is complicated. I go back home around midnight three times a week. We must try to cope with it" (Researcher, biology).

– "This is not an ideal situation for researchers, because of our strong academic background, we like to take our time to do things and appropriate them. Once we have entered a system of deliverables in the short and medium term, on tight objectives because a client pays us to do so, the research relationship is a little different" (Researcher, chemistry).

Box 3.6. *Illustrations of temporal slack reduction effects: testimonials from Druglab researchers (source: Bobadilla [BOB 14])*

3.3.4.3. *Conflicts between professional and personal life*

Beyond the negative effects on creativity, increase in workload due to the acceleration and frequency of temporal slack reduction also generates tensions with respect to personal life. The time that is consumed by a role (work) is not available for other roles [DER 08].

The concern today is that extreme time pressure reflects change through the temporal boundaries of life at work. The continuation of work in extreme time constraint requires the continuation of professional activities outside the schedules initially defined, which constitutes a risk to the physical and psychological health of individuals.

3.3.5. *Spatial slack reduction effects*

Among the various resources available within an organization, space has been theoretically and empirically underestimated; meanwhile, it is a key resource. Thus, we know very little about the impact of space slack reduction on creativity and innovation. However, recently, the workspaces and creative spaces have changed drastically inside and outside the organization. Open innovation, fab labs and third places have come to transform the landscape and space where creativity and innovation take place. Inside, *open spaces* have become established in the R&D universe, with a view to reducing costs while hoping that the idea will promote creativity. Thus, changes in space focus on the instrumental characteristics of space by exploring the relationship between organizational space and goal attainment without addressing physical space reduction effects on creativity. Space can not only be conceived (for instrumental use), but also “lived” [LEF 91] and understood.

3.3.5.1. *Difficulties with open spaces*

The most appropriate workspaces depend on the stages of the creative process and culture of the organization. Thus, it is important to have spaces that encourage isolation in the first phases of idea generation: quiet work is essential for literature reviews, reading and testing. R&D professionals indicate how they prefer closed workspaces because it encourages concentration.

On the contrary, for idea sharing and evaluation phases, semi-open spaces and reception lounges seem to be important.

Other studies have also shown that open spaces lead to a decrease in communication compared to private offices [HAT 90] as well as a decrease in performance and motivation. Allen [ALL 77] pointed out how communication and innovation in R&D organizations could be enhanced by desk approaches that stimulate informal interaction. For example, partitioning between two workstations in an open space creates a remoteness related to the suppression of visual connection. He also discovered that the frequency of worker interactions in an R&D center decreased exponentially with the distance between their offices – an effect known as the “Allen Curve”. Paradoxically, being physically closer in an open space does not have a direct impact on the frequency of communication. Instead, individuals and teams may tend to talk less easily for fear of disturbing the others [BOB 14]. Even when they were in the same building, researchers from different floors almost never interacted informally. Thus, we observe that space may or may not encourage interaction, depending on how it balances the three dimensions: proximity, confidentiality and permission.

– Proximity: people often assume that proximity is purely based on physical spatial factors: to what extent are employees close to one another in physical distance. Distance is important but it is not only the physical attributes of a space that influence informal interactions. "Proximity", as we use it, depends on traffic models that are shaped just as much by social and psychological aspects. Allen [ALL 77] and Bobadilla [BOB 14] indicate that in order to improve the dissemination and sharing of ideas, it is important to create spaces with many shared resources (coffee space, relaxation break, shared printer spaces). The social geography of a space is a crucial element of its physical layout.

– Confidentiality: physical privacy requirements are most obvious. People must at least have a space where they can interact without being heard. To ensure this confidentiality, spaces must be conceived taking into account visibility and acoustics; confidentiality is enhanced when others cannot see the person we are talking to and when we can see others approaching or within earshot. There is a subtle implication here: genuine privacy allows us to control others' access to us so that we can choose whether to interact or not. Although this may seem counterintuitive, research shows that informal interactions will not fully develop if people cannot avoid interacting when they want.

– Permission: the social dimension of permission is more obvious than the physical dimension, but both are critical. Culture and convention influence our vision of what constitutes an appropriate behavior in a particular environment. Sometimes, artifacts in a space strongly affect its social designation. For example, it is possible to introduce a recreational area within the workspace, but in order not to perceive this negatively, it is necessary that everyone adheres and that in the organizational culture, play during work be accepted and dealt with according to social conventions.

3.3.5.2. *Perceptions of space as symbolic mirrors of the environment and culture*

Workspaces and office layouts are just the facade of what spaces reflect: social interactions, organizational climate and culture. The idea is that the instrumental function seems to reflect more accurately an in-depth manifestation of culture and builds tangible evidence of the work climate and environment within an organization.

Thus, we observed that the descriptions of physical space made by individuals reflected the effects of slack reductions and the climate of the organization. For example, feelings of "partition", "division between floors", "distance between teams" and "spaces that do not promote interaction" are related to physical space, as well as reflecting the culture and climate of the organization (see box below). Thus, space is very important not only for creative work, but also for its symbolic function.

– "The relationships have not grown stronger, people are rather individualistic. With open spaces there is no difference, gathering people. Some colleagues have really suffered from the PSE, they have difficulties moving on. Presently, they are still bitter about this situation" (Research Associate, Pharmacology).

– "What is strange is that the reorganization was made to break down barriers a little and this is not necessarily what was created, it's a shame. We have chemistry on the ground floor, biology and DMPK on the first floor, and ultimately, apart from meeting from time to time during joint meetings, there are not so many exchanges as such. It is not voluntary. Normally the reorganization had to unite us, but it was not the case. It was rather a trauma, everyone kept to oneself" (Researcher, biology).

– "We relocated. Simultaneously with the relocation that took place the same year, there was the social plan. We changed not only to a reduced human environment but a building environment. We said, to an extent, it will do us good. Did you see this building that was initially planned for 120 people? We are more than 70. With these open spaces! How awful!"(Researcher, chemistry).

"In an office where there is so much of noise, no. For me, this is not the place to reflect. Sometimes ideas come and if there is noise at that time, it is very easy to lose the train of thought. When I feel that I need to think about some things, the open space is not ideal. You need to take your material and go elsewhere, to find a little calm and quiet" (Researcher, biology).

– "Before it was more natural. We would meet in a small office and start discussing on a subject. Here, if we want to discuss a topic, either everyone hears it and it can disturb people, or we must go to a room and this makes it less natural" (Researcher, DMPK).

– "Absolutely no exchanges. It's very complicated. Very paradoxically, with my 2 neighbors, we are as close as you and I, 30 cm apart. And we can spend an entire morning without seeing each other because there is a small partition between us" (Researcher, biology).

Box 3.7. *Some illustrations of the relationship between physical space: spatial slack reductions*

3.3.5.3. *Importance of creating moments and sharing spaces*

After human slack reductions, the creation of spaces and moments to address the process (the way events were experienced, different reactions), heal wounds and leave the past behind is important. Emotional and cognitive responses must be taken into account during these processes. To do this, issues such as mindset and mental stress must be addressed, and work that needs to be done by everyone in the organization in order to recover from the event as a group has to be taken into account as well. When changing the physical space, organizations need to consider the importance of spaces that stimulate informal interaction, as well as spaces dedicated to private and collective thinking. The teams themselves can recreate spaces for social activities (sports, festive events, conferences) that facilitate the team's cohesiveness and focus of attention on other topics. Outdoor activities, outside the organization, seminars and brainstorming sessions are important, as is the separation between work and personal life.

3.4. Mechanisms linking slack reduction and creativity

R&D slack reduction – whether human, financial, temporal or spatial slack – affects creative processes in their different stages (not only generation but also sharing, filtering and evaluation of the ideas). This link is due to slack reduction effects on the following three key mechanisms in creative processes: focus of attention, ability to "travel" over time and support from leaders.

3.4.1. *Focus of attention*

Individual approaches to creativity show how the creative individual has the ability to evolve between opposing poles, such as focus and defocus of attention, play and discipline, passion and objectivity, disorder and order or freedom and control.

Focus of attention is defined as narrowing the level of attention: more attention is given to central signals, while peripheral signals are neglected [COH 78, GEE 76]. As a result, focus of attention improves the performance of an activity if it requires only central benchmarks, but it compromises that of an activity in which the perception of peripheral signals is necessary. However, the ability to focus and relocate attention is crucial for creativity

[CSI 96]. In other words, Amabile [AMA 96] emphasizes how creative cognition involves transition between a divergent and convergent mode of thinking.

The frequency of slack reductions discourages individuals and teams from moving away from attention targets. For example, with reductions in staffing and financial resources, the shortening of time available for research or increase in supervision and requirements for filtering and evaluating ideas compel teams to be "hyper-targeted". This narrowing of attention disrupts creativity.

Longitudinal research on the effects of downsizing has shown that they reduce over time and are, ultimately, relatively short-lived [AMA 99, SUS 07, MEL 10]. However, Bobadilla [BOB 14] showed that the frequency of slack reductions no longer provides a way out of the focus of attention. The problems caused by the frequency of downsizing, difficulties in implementing them and deterioration of inter-individual relations generate conflicts of attention [BAR 86]. Poor implementation of downsizings generate conflicts of attention and fixation by destroying the possibilities of timelessness, to the extent of becoming sources of conflicts between work and personal life.

This "hyper-targeted" attention seems to be inappropriate, especially with regard to the generation and sharing of ideas, as well as the intrinsic motivation, which constitute the source of a team's creativity. A creative effort without intrinsic motivation is almost impossible. The damage caused by slack reduction frequency is related to a narrowing focus of attention, a decrease in cognitive availability for creative work and a shift to explicit knowledge.

3.4.2. *Ability to "travel through time"*

R&D work is unique in that it establishes original temporal relationships [YLI 09]. To illustrate, in the creation phase, it is essential to not only rely on the past (knowledge and experience) but also project oneself into the future (open the mind to new horizons). In innovation processes, R&D teams perceive and interpret time, use it and allocate it in different of ways. For example, R&D teams often encounter a changing as well as frenetic pace.

Slack reductions symbolically and explicitly interrupt the temporal patterns of teams and organizations over time. Symbolically, in the short term, these reductions lead to an interruption, a temporary break, in the history of an individual or team, thus questioning his/her past, present and future identity. Explicitly, if we talk about reducing human slack, the activity stops because the uncertainty created by the PSE induces people to stop all activity. The human and temporal slack reductions affect the cognitive availability of creative workers, destroying the possibilities of timelessness. Timelessness indeed requires the creation of a psychological and sometimes physical space in which teams can become involved in the creative task, far from worries, problems or distractions. Symbolically, the future is of particular importance for R&D teams and inadequate implementation of slack reduction anchors teams in the past, thus focusing their attention on past and negative events.

Over time, the different slack reductions constrict the present time by narrowing the focus of attention. This means that teams are forced to focus on the short term in order to cope with the pressures of their environment. Bobadilla [BOB 14] highlighted that creativity is related to how teams can travel through time (past, present and future) and transform their memories of the past and their vision of the future to open up, in the present, new opportunities for creative action.

3.4.3. *Support provided by the leader*

R&D team leaders seem capable of influencing or even supporting their team in this "time travel" process. The role of the leader is to project a vision that helps teams to look toward the future.

In the slack reduction phases, the leader's support is very important, at least if the teams trust him. Mishra and Spreitzer [MIS 98] emphasized the role of trust in downsizing management, conceptualized by Rousseau and Tijoriwala [ROU 98] as "a psychological state with the intention to accept the vulnerability of the other person according to the positive expectations, intentions or behaviors of another". There are three types of trust, namely emotional, cognitive and symbolic.

– Emotional trust is demonstrated when team members trust their leader; they sympathize with his/her image and feel that he/she creates an atmosphere of psychological security.

– Cognitive trust is based on a member's perception of the reliability and competence of his/her leader to perform the technical part of the work.

– Symbolic trust is based on the knowledge of others. It is the trust that results from the conviction that the leader gives meaning to, the confidence of being safe to go from A to B (travel through time), with this leader.

In the specific context of R&D slack reduction, trust in the leader plays a key role, because it encourages risk taking that is inherent to creativity. Symbolically, in R&D slack reduction processes, the leader is considered reliable when he/she is able to connect the present and future of the teams. This means that it can provide a clear vision of where research should be directed to. It is precisely because of the uncertainty created by slack reductions and the very nature of the research activity that the role of the leader is essential in providing a vision of the research and structuring the activity of his/her teams.

Thus, leaders can play an important role in these slack reduction phases, but this raises the question of their leadership skills. In fact, their role in these slack reduction phases calls for empathy, a strong ability to work in a context of uncertainty and tolerance of ambiguity. R&D leaders have scientific and technical backgrounds, and they need additional training in management, especially to be able to support their teams in these slack reduction phases.

It is also important for organizations to appoint as downsizing leaders people who can help teams refocus attention and "travel through time".

3.5. Conclusion

Slack reductions are based on several interrelated dimensions. Human and financial slacks thus appear as parent or major slack forms with strong influence on other types of slack reduction (spatial and temporal). In addition, human slack reduction effects permeate the spatial slack of those creating. In the same way, financial and human slack reductions affect temporal slack. Therefore, the apprehension of slack reduction as a reduction of various types of slack (human, financial, temporal and spatial) facilitates a better understanding of these phenomena.

In this chapter, we also emphasized the fact that the frequency of slack reductions generates adverse effects to creativity. R&D managers must know the slack forms behind the human or financial slack that are relevant to creativity. While they have little leeway in terms of downsizing human and financial resources, it is important that they concentrate more on how to redeploy temporal and spatial slack.

One must consider not only the quantity of these slack resources but also the quality. R&D teams can create within scarcity environments, but organizations need to protect the quality of these slack resources and avoid their degradation over time. The symbolic aspects should not be neglected, whereas the biggest managerial challenges facing R&D organizations currently are no longer technical, but rather how to instill more "soul", "spirit" and hope in the work environment of the creative teams.

4

Managing R&D Professionals: HRM Practices and Current Challenges

Since the 2000s, managing R&D professionals has been the subject of increased theoretical and empirical interest as companies have engaged numerous HR management initiatives destined for R&D communities. Despite many HRM practices being developed and numerous debates in academic research on these issues, there remain a great number of challenges for the Human Resources function in R&D [CHA 12]. In section 4.1, we highlight the relatively recent and structurally complex nature of the relationship between HRM and R&D. We then review the different HRM practices (strategic HR planning, recruitment, assignment and mobility, evaluation, remuneration, careers, competence management) as observed in contemporary R&D organizations. Finally, considering the profound changes witnessed by R&D most recently, the limits of certain HRM practices are reported, calling for an in-depth reinvention of at least some of the management methods destined for R&D professionals.

4.1. HRM and R&D: complex relationships

Although for a long time the R&D world remained impervious to the interventions of Human Resources Departments (HRD), this is no longer the case. Over the past two decades, HRM practices have become increasingly significant in this scientific and technical world. This generates a number of tensions. These tensions are partly due to the mutual lack of understanding between these two worlds, and also to the underlying fundamental differences in their ways of thinking (section 4.1.1). Thus, one of the areas of structural tension relates to the dilemma between standardization and differentiation of HRM practices deployed in R&D compared to the "usual" practices applied to other employee groups of the company (section 4.1.2). The rise and the increased popularity of project organization in R&D is one of the key elements that challenges the relevance of traditional HRMs practices and calls for the invention of new methods of regulating work in R&D (section 4.1.3).

4.1.1. *R&D: a world that has long remained foreign to HRM regulations*

R&D has long been managed according to the endogenous rules of a professional world marked by the very strong influence of peers and epistemic communities that extend beyond and across organizational boundaries. Until the early 1990s (in France, for instance), it was normal that R&D, and especially the "R" part within it, should be self-managed according to its own logic and without the intrusion of "external" actors having a managerial vision. To this day, recruitment, individual evaluation and career management practices are marked by the influence of reputation phenomena within the community and by judgment criteria that are very similar to those of the academic world (such as the expertise, originality and robustness of the knowledge generated).

This industrial R&D world has always been very porous with regard to academic research. Industrial R&D researchers and engineers who are trained and socialized within the higher education and research system, once hired in a company, continue to maintain close links with the academic world within the framework of their professional activities. A certain mimetism in job titles or promotion criteria gives industrial R&D professionals legitimacy with regard to academic partners and public authorities (see Box 4.1). This legitimacy enables them to establish a relationship with academic research and to reap a good number of benefits from these links, such as opening up access to knowledge and ideas, equipment, recruitment pools, funding, etc.

Rhodia was a French specialty chemicals company, acquired by the Belgian chemicals group Solvay in 2011. Until the 1990s, the structure of the R&D centers of Rhodia was based on the classical divisions of the field, with a separation between the organic and inorganic (mineral) chemistry departments. Within these centers, the laboratories were modeled according to the professions of the chemical industry: synthesis, processing, application and analysis. These subdivisions, as well as the symbolic hierarchies between these professions, corresponded to those of academic research. This mirror-like structure facilitated the legibility of the company's activities for academic partners. Thus, when the Rhodia mineral synthesis laboratory referred to the mineral synthesis laboratory of a particular university or research organization, the researchers who established a relationship had very similar profiles and were working on the same topics. It also enhanced recruitment because young doctors from an organic synthesis laboratory who had a PhD degree in this field could easily determine which industrial laboratory to contact to seek for a job. Each industrial laboratory therefore registered within a well-identified academic community, hosting conferences focused on its field, with journals specific to this area. Industrial researchers' careers were also simplified, since they prepared and presented an HDR (accreditation to supervise research) in their scientific field recognized by the academic world. They were then entrusted with the supervision of doctoral students enrolled in this field. Mimetism also concerned job titles ("Head of Laboratory", "Research Director", etc.) and promotion criteria. This fostered the R&D department's ability to establish relationships with the academic world, identified as an essential partner, and to harness and exploit resources from it.

In the 1990s, the primary concern of R&D management became the increase and acceleration of the transformation of research work conducted by R&D into concrete products and services. To encourage this, a closer link was established with the business entities that became the main clients and sponsors of R&D. This was equally reflected in an evolution of the structures of R&D centers henceforth divided according to these internal "clients" and the fields of application of chemicals (building, automotive, electronics, etc.).

Box 4.1. *Mimetism with regard to academic research: the case of Rhodia (source: Gastaldi [GAS 07])*

This near autonomous regulation of R&D, which was fairly widespread until the 1990s, was not specific to R&D. Thus, organizations had quite similar operation methods and hired highly qualified staff (teachers, doctors, magistrates, etc.), which the sociological literature referred to as "professionals". In these organizations, HRM, as stated by Pichault and

Nizet [PIC 13], was of conventionalist (or deliberative) type: admission and professional assessments were collegial; recruitment and promotion decision-making processes were subject to co-optation by a peer panel; training was fully controlled by professionals.

However, though R&D is not the sole professional field governed by this type of HRM logic and these types of practices, it stands out within large industrial companies. The most widely disseminated HRM practices were indeed designed in large corporations, but for functions other than R&D and estranged from these professional communities. Therefore, it is no exaggeration to say that HRM and R&D share a mutual lack of understanding.

For HRM specialists, R&D activities and technical expertise are still largely a *terra incognita*. This stems from cognitive complexity and cultural distance. It is indeed very difficult for the managers to understand such a complex activity (with specific vocabulary and concepts) which is moreover intangible. The R&D activities are in fact largely inaccessible for anyone outside the field. This could even be the case between the scientists themselves: a polymer chemist, for example, who is unable to understand and evaluate the quality of the work of a mathematician specializing in specific types of algorithms. This is what justifies the relevance of the peer review. Therefore, we can imagine how it is even more difficult for actors in the HR function, mainly from "non-scientific" training (their background being generally human and social sciences: law, psychology and particularly management). In addition, the vast majority of them have neither prepared nor defended a PhD thesis nor been trained in research. Beyond cognitive distance, these communities are also alienated by cultural distance. Thus, HR managers perceive researchers as people with strange and unusual ways of working, rebellious to any management regulation. This cultural distance is less significant for engineers working in the "Development" part of R&D and engaged in more "concrete" activities (in the sense of greater materiality and closer proximity with applications), but it still remains important.

Symmetrically, from the point of view of R&D personnel, HRM, just like management in general, is perceived as a set of administrative activities, without much interest. R&D engineers and researchers often have a very inconsistent understanding of what HRM is all about. This is often limited in their minds to the payroll or other administrative tasks. In addition, if it is explained to them or if they observe that the HR function has a potentially broader scope of intervention, the discussion turns to the competence, or incompetence, of HRM specialists. Since they do not have a scientific and technical training background, what could be their contribution? How can they be relevant and bring added value to decision-making with regard to recruitment or promotion, for example? What is their legitimacy to intervene in the operational modes of a professional world of which they understand nothing? Therefore, any attempt to penetrate one's territory is viewed with some suspicion.

In a bid to bring these actors closer together, the company initiates training and communication actions. Thus, training introduces R&D professionals to the mysteries of management. Cooperation is sometimes organized around one-off events, particularly during meetings with higher education institutions aimed at recruitment. We might think that through constant collaboration, the distance between the two worlds should scale down. However, this is not enough. On the contrary even, the more we know each other, the more we realize that we are not driven by the same logic and that these logics are neither spontaneous nor necessarily reconcilable [BOB 15]. The distance is particularly wide between the scientific and technical logic advocated by researchers and the managerial logic that HR departments intend to promote. While the former, discouraged by the routine and normative constraints that limit their creativity, aspire to freely produce an original work, the latter intend to make the same managerial order prevail in the entire company.

Moreover, would it not be uncommon for relationships between R&D and HRM to take on the proportions of a standoff, when they do not escalate into confrontation. What may have changed over time is the powerful upswing of HRM, as well as economic and managerial logics in general that take precedence over the magnitude of scientific and technical logic, hitherto very strong in large French industrial groups long managed by former engineering school graduates.

Thus, HRM now receives greater attention in R&D, but relationships with scientific and technical professionals are still complicated. Although the latter partially, and in spite of themselves, integrate management logic that has become very significant, including in R&D, they consider the HR function as the linchpin behind a broader managerial rationalization project (see Chapter 3) combining control, economic logic and bureaucratization, which is indeed what it is. The issue then is the ability of the HR function to be something more than just that. Could it be also a source of added value through the actions it undertakes within the R&D field? This is an essential discussion that will be continued throughout this chapter.

4.1.2. *Recurrent tension between standardization and differentiation*

Once HRM penetrates the R&D world, the issue of the application of "standard" practices (similar to those deployed in other functions), or, in contrast, "differentiated" practices (adapted to R&D specificities), will arise. The standardization-differentiation dilemma exists for other communities, but is particularly intense in R&D. Indeed, it combines a good number of specificities, which distinguishes it from the other functions of the company: the R&D activities are creative, uncertain, complex, singular, intellectual and partly immaterial, open to the outside, etc. (see also Chapter 2). They are driven by highly trained and autonomous professionals, invested in epistemic communities. This standardization-differentiation dilemma of HRM practices in R&D has been the subject of several studies [CHA 05, DHI 08].

Each alternative has certain advantages and disadvantages. Standardization simplifies the application of HRM processes by providing a common base, and it enables the specialization of the work of HR actors and thus a rise in professionalism of the latter (and incidentally a rationalization of the costs of the HR function!). The deployment of homogeneous practices throughout the company is also a source of internal consistency that is put forth to encourage collaboration and crossfunction mobility. It addresses the concern for equal treatment between all employees of the company. It helps the HRD to increase its influence and assert its central role, managing all processes covering the entire company from the head office. However, standardization widens the distance between the HR function and R&D, because of the negation of the specific nature of R&D and its actual conditions for operation. The R&D professionals could feel disqualified when obliged to follow the standards imposed by organizational actors whose legitimacy they question.

Conversely, differentiation makes it possible to adapt to the specificities of R&D activities and professionals, and to their cultural particularities. The implementation of HR practices specifically designed for R&D communities makes them more legitimate, thus facilitating their appropriation by R&D stakeholders (managers and professionals). Another positive effect of the differentiation of HRM is the possibility to assert its expertise in relation to that of the technical world. However, differentiation raises several questions. It indeed introduces a violation of equal treatment. It can introduce additional impediments relative to the possibilities of crossfunction mobility for R&D communities, in the context where R&D employees are already perceived as atypical and not specifically bound to move from one department to another.

Thus, the appropriate question, rather, is to know where to position the cursor on the continuum of situations moving from total differentiation to absolute standardization: differentiating or homogenizing to what extent? Which HR variables (recruitment, remuneration, training, mobility, career management, promotion of professional equality, etc.) should rather be differentiated or standardized? If we want to differentiate HRM practices, on what grounds will they be actually differentiated and how? In any case, before following a certain option, it is necessary to question the reasons behind such action, while avoiding to conform to dominant practices by a mimetic behavior of copying the practices of other companies without questioning what relevance they would be of to us. It is equally important to ponder on the actions that can help us to avoid or limit the adverse effects of the selected option, since, as previously indicated, standardization and differentiation each have certain disadvantages.

Many companies are trying to adopt a middle-of-the-road approach, associating some degree of standardization (following a managerial rationalization process) and partial adaptation to R&D specificities. Several factors promote a greater homogenization of HRM practices between R&D and other functions, and this coupled with a convergence towards HR models that are globally disseminated, especially the individualizing model (as referred to by Pichault and Nizet [PIC 13]). Here, the internationalization of large industrial undertakings and their R&D (see Chapter 1) operate in this direction. However, studies suggest that there is not necessarily a strong alignment as we might have thought.

As for the convergence towards a global and single model, Béret *et al.* [BÉR 03] showed the persistence of national differences. Thus, recruitment and remuneration systems continued to be designed at the national level, reflecting the legal and institutional differences of countries that are taken into account in human resource management practices applied in the R&D sphere. Further studies have to be conducted and considered in this perspective, given the recent intensification of phenomena around the internationalization of R&D and globalization of management models.

Regarding the convergence towards the individualizing model, Dhifallah *et al.* [DHI 08] demonstrate on two cases, Saint-Gobain and STMicroelectronics, that the latter has indeed gained considerable ground in R&D. However, they point out how the conventionalist (or deliberative) model practices persist or at least how hybridization occurs through an adaptation of individualizing practices to R&D specificities (for example, an individualized remuneration indexed to patent filings).

Before describing the HRM practices most frequently observed in R&D today in detail (in section 4.2), it would be important to highlight how changes in the organization of R&D have enhanced the development of specific HRM practices. Thus, the intensification of project organization since the late 1980s called for a reflection on HRM methods adapted to these particular forms (temporary, matrix, multidisciplinary) of work organization.

4.1.3. *Project organization: a necessary source of adaptation of HRM in R&D*

Project organization, which is now common in R&D activities (see Chapter 2), suggests the recognition of the need for specific HRM practices. The latter are a combination of recognition, on the one hand, of project mode specificities, which are of course not reserved for R&D and can now concern a wide range of business activities, and R&D specificities, on the other hand.

In so far as the quality of human resources is a prerequisite for project performance, especially R&D, and where the project is itself a vector of skills development[1], HRM has to give attention to the management of people involved in projects. Its traditional practices were destabilized because of projects' characteristics disrupting the classical (hierarchical, vertical and

1 In the beautiful words of Jolivet [JOL 03, p. 201]: "*L'homme fait le projet qui fait l'homme*" (the man carries out the project that makes the man).

functional) organization within which and according to which HRM was built. Indeed:

– projects involve, to varying extents, dual project/profession authority;

– they question the authority of the professions, embodied in classical functional structures.

Studies then focused on the issues raised by this new organizational form and its potential impact on HRM, as well as on the relevant responses that the latter could provide to these specific situations of temporary assignment of individuals to cross-functional projects. These include Baron [BAR 99], Paraponaris [PAR 00], Garel *et al.* [GAR 03b], Zannad [ZAN 08], and Loufrani-Fedida [LOU 11].

As Garel *et al.* [GAR 03b] have shown, the project structure raises multiple issues for HRM, including the recruitment of project managers, training, evaluation, remuneration and career management. These issues are discussed in the next section, as the different HRM practices currently observed in R&D are reviewed.

4.2. HRM development in R&D today

In the following sections, we propose an assessment of the most common HRM practices in R&D today. We rely on our own observations [GAS 06, GAS 16, CHA 12, LEL 14] and academic literature. We seek to uncover the emergence, objectives and concrete realization of the major HRM practices as applied to the management of R&D professionals: strategic HR planning, recruitment, assignment and mobility, evaluation, remuneration, careers, competence management, etc.

4.2.1. *Strategic HR planning*

HRM should not only react to the needs and difficulties encountered as and when they arise. It is important, although difficult, that it tries to anticipate their occurrence, in order to have the necessary time to implement appropriate actions that help, to the greatest possible extent, to avoid situations of mismatch between the organization's needs (specifically in terms of staffing needs and aspired competences) and the actually available resources. Efforts are being made towards applying a strategic planning approach for R&D activities. Such an approach mostly concerns, large companies, however, as the smaller ones have more difficulties in anticipating their future. Large companies have embarked on major projects in order to anticipate future critical competencies. Orange, for example, tries to predict what impact would environmental, strategic, organizational, as well as technological choices of the company have on the R&D professions (in terms of staffing needs and required competences). This work is carried out by both HR departments and R&D management.

Strategic HR planning aims to identify the jobs and competencies to be strengthened, developed or upgraded, as well as those whose downsizing has to be prepared. The result of this is differing decisions that mobilize the whole range of HRM tools. When there is a need to increase the number of employees or to develop new competencies, in anticipation of the expansion of a market, successful innovation and the rise of certain technologies (for example, digital technology, Big Data or artificial intelligence), various HR practices will be implemented: recruitment, incoming mobilities, training, etc. Other decisions can be taken that do not fall within the scope of HRM. Such is the case with external growth transactions, when company takeovers make it possible to appropriate new product portfolios, markets, patents and R&D teams. The same is true regarding the use of outsourcing and service providers who offer particular expertise. Indeed, business service providers are growing, and some companies such as Altran Technologies are offering engineering and innovation solutions.

Companies may also face more difficult configurations, where staffing and skills needs are declining, which requires downward adjustments of existing resources. This is the case when products are abandoned, markets divested and expertise outsourced or when technologies become obsolete. HRM then has to initiate actions (outgoing mobilities, new skills training, VAE, voluntary departures, etc.) as soon as possible to prepare re-qualification and re-training for the concerned employees and to accompany the possible departures.

However, deploying a truly strategic planning is a difficult objective, regardless of the company's functions and professions, but perhaps even more so in the area of R&D. Thus, in an uncertain environment that is subject to rapid changes, instrumental approaches are often unable to predict possible market upheavals as well as future technological advancements (internally and externally), including the opportunities they can create in terms of innovation.

4.2.2. Recruitment and integration

4.2.2.1. *Recruitment adapted to the specificity and diversity of the R&D profiles*

Recruitment needs are derived from a variety of factors. They can emanate, as pointed out by Akhilesh [AKH 14], from the diversification of activities, creation of a new product line, development or integration of a new technology, establishment of a new partnership or any other form of collaboration. Recruitment needs also depend on staff turnover in R&D, since R&D professionals can quit their positions for many reasons: promotion within R&D, resignations, dismissals and moving to other functions of the company. Indeed, crossfunction mobilities are strongly observed today in large industrial companies; they occur earlier in the professional career and concern almost exclusively outgoing mobilities from R&D towards other functions. Reverse mobility turns out to be very rare. The latter is not particularly beneficial for individuals and is also seldom desired by companies that generally choose to recruit externally when it comes to R&D.

Beyond the fact that it is mainly oriented towards the external labor market, the R&D recruitment process has other characteristics resulting from the specific nature of R&D activities and the particularity of candidate profiles.

A strategic dimension of recruitment management lies in the identification of the right profiles sufficiently in advance. Within a context of "war for talent", this involves targeting and attracting potential candidates before they are identified by competitors. Thus, establishing relationships with engineering schools, universities and research organizations is very important. The position of "campus manager" has thus been created within HR departments in order to identify, attract and pre-screen young people with interesting profiles while they are still pursuing their studies. HR and R&D managers are in charge of identifying national and international higher education and research institutions, which constitute relevant recruitment pools for the company as they provide specialized training or conduct relevant research activities. They must then communicate with the institutions about the company's businesses, career opportunities, the recruitment process, etc. Beyond the communication efforts, various actions could reinforce the link between the corporate world and the academic institutions: internships offered to students, scholarships, joint supervision of research papers and PhD theses, various grants, post-doctoral internships and, more recently, teaching and research partnerships, participation in the development of academic curriculum, etc. Such actions help to adjust the content of academic learning with companies' future needs, which will be mutually beneficial for students and companies.

The recruitment needs and the profiles of potential candidates could vary depending on the business sector, the scientific and technological field and the innovation strategy of the company. A company oriented towards breakthrough innovation will prefer PhD holders, and more particularly those with postdoctoral experience, while a company that prioritizes incremental innovation may prefer profiles with a lower level of academic training. For the companies that are targeting the candidates with doctoral degrees, it could be very beneficial to invite PhD students to complete their thesis within the R&D laboratories. Beyond the possibility to deepen a potentially innovative topic, such a practice could also serve as a pre-recruitment process. In France, a specific agreement – CIFRE convention – could help the companies to reduce the cost of hiring of a PhD student.

In France, the CIFRE convention, *Convention Industrielle de Formation par la Recherche* (Industrial Agreement for Training through Research), is a public scheme that aims at fostering the development of public–private partnership research, by placing doctoral candidates in conditions that would lead to employment and by promoting their future professional integration. It subsidizes any organization under French law (companies, associations, territorial authorities, etc.) that hires a doctoral student as part of research collaboration with a public laboratory. There is no age or nationality requirement, and all disciplines are involved. Works carried out must normally result in PhD thesis defense within three years. CIFRE conventions are fully funded by the French Ministry of Higher Education and Research, which has entrusted their implementation to the *Association Nationale de la Recherche et de la Technologie* (National Association of Research and Technology) (ANRT).

This scheme is based on the collaboration of four actors: the company that recruits a doctoral candidate (on a permanent contract or a fixed-term contract basis) and entrusts the candidate with a research mission; the academic research laboratory that supervises the work of the doctoral employee who is enrolled in a doctoral program in the graduate school to which the laboratory is affiliated; the doctoral student who devotes 100% of his or her time to research work and shares it between the company and the research laboratory; the funder, ANRT, which contracts with the company and pays the candidate a subsidy.

Box 4.2. *What is a CIFRE convention? (source: http://www.anrt.asso.fr/fr/espace_cifre/accueil.jsp)*

Though most R&D recruitments concern young researchers and engineers who have recently completed their training, the company may sometimes want to integrate more experienced profiles. These researchers often come from partner companies, clients or competitors. They are generally well acquainted with the business sector, have valuable networks and may have had project management or hierarchical management experience, which could be beneficial for a company that is willing to recruit senior R&D staff members.

Another characteristic of the recent period is the ever-increasing requirements concerning the level of competencies expected of new recruits. Beyond technical and scientific skills that remain essential, the so-called "behavioral" skills are becoming increasingly significant in recruiters' evaluation grids. Indeed, it is expected that candidates should demonstrate some level of curiosity, leadership, ability to communicate and convince, entrepreneurial spirit, etc. [CHA 12]. Companies are becoming more and more demanding and could even paint an (overly) ideal picture of an R&D employee who must excel in a large number of fields and be capable of taking on various roles. The risk would be the inability to fill the positions because of highly selective criteria that greatly reduce the potential recruitment pool.

These difficulties are exacerbated as the evolution of R&D management methods (see Chapters 2 and 3) comes with contradictory expectations. This is the case when the company asks R&D professionals to let go of routines, be creative, invent the future, and also comply with work processes and more formalized management rules. Holmqvist and Spicer [HOL 12] point out that the companies try to create "ambidextrous employees" and Chapters 2 and 3 have pointed out the tensions that these paradoxical injunctions generate in the daily work of R&D professionals.

In view of the accumulation of skills and roles expected of R&D professionals, it would be more relevant to reason in terms of not only individual but also collective competencies. This would make it possible to more reasonably consider how all the expected skills and roles can be assumed by the team as a whole, and not by each of its members (see Chapter 5). Beyond the complementarity of skills and roles, the compatibility between the new recruit and the team he/she will integrate should also be taken into consideration. However, we must also look beyond the position that will be occupied in the short term, because the recruitment of R&D professionals is always done in a longer term horizon. It then becomes necessary to question the candidate's career aspirations as well as their professional development abilities, particularly towards expertise, project coordination, hierarchical management or other functions.

Alongside the rise of these new expectations (behavioral skills, ambidexterity, potential for development, etc.), we observe a much stronger presence (and power) of HR actors in recruitment processes, where previously they could be completely excluded. HR actors are better positioned to appreciate candidates' soft skills, their motivation, career aspirations and ability to integrate and grow within the company, including outside R&D, in connection with the possible paths and those desired by the company. On the other hand, because of the highly specialized nature of scientific and technical skills, peers remain involved in the sourcing (detection) and selection of candidates, alongside HR managers and the direct hierarchy. Such a heterogeneity of evaluators brings more detailed assessment of the different facets of candidates' skills. It also enables them to better picture what their future work environment could look like in the short and medium terms. It is indeed important to pass a carefully conceived message to candidates, in order to promote a realistic vision concerning the content of their future work, missions and projects they will be working on, career opportunities, the company's operating rules, etc. This helps us to prevent the new recruits from having any disappointments when they actually enter the company.

4.2.2.2. *The crucial phase of integration*

Recruitment does not end with contract signing and the integration phase represents a crucial step that will ultimately determine recruitment success or failure. Integration is a socialization process through which values and standards are transmitted and professional identity is created. Despite the importance of integration, this crucial process could be missing or not efficiently managed in organizations.

The first year within an R&D department is often a challenging experience and a source of stress [KAT 04] for a new recruit. This involves trying to be accepted by the team and peers, becoming acquainted with the company's operating rules and feeling useful by deploying his/her skills to the service of the company's strategic activities. If this phase is not successful, the risk of constrained or voluntary departure of the employee increases considerably.

It is therefore important to reflect on both the content and the process of socialization [KAT 04] in order to deploy practices that enhance it. Regarding the content, explaining acceptable and required social attitudes in an R&D environment will facilitate the integration of new employees. This includes supporting them in developing their roles and organizational identity, while allowing them to gradually feel at ease in carrying out their work, from the understanding of procedures and communication channels to the mastery of the technical dimensions of the profession. As for the socialization process, it depends on the quality of interactions of the new employees with the other actors of the company. This process could be more or less long depending on organizational factors and personal characteristics. The success or failure of the integration phase has a considerable impact on the employees and could potentially impact the rest of their professional life. It is therefore necessary to create an environment that is conducive to the new recruits' initial stages within the company. Raising awareness among local managers, as well as the implementation of specific training programs, could be among the possible actions facilitating the integration process within an organization.

4.2.3. *Assignments and mobility*

4.2.3.1. *Assignments to projects and selection of project-actors*

With the growing importance of project organization in R&D, the assignment of diverse project contributors becomes a crucial issue. The success of the project depends largely on their competencies and their investment. However, these assignments are also critical for R&D professionals themselves because, beyond the immediate consequences, it is their subsequent career trajectory that can be greatly influenced by the visibility of projects to which they are assigned, by the degree of learning generated (or not) by the work on the project, and, of course, by the project's success or failure.

While the assignment choices are usually made by managers, it is important that HR considerations are also taken into account. It is crucial to assure the fit between the employee profile and the project, in terms of both hard skills (scientific and technical) and soft skills (ability to communicate, to adapt to other professional logics, to understand clients' needs, etc.). The company should pursue the assignment policy that creates learning opportunities and avoids the risks of skills obsolescence. Thus, a succession of projects in "exploitation" (or operations) poses risks of exhaustion of expertise and skills obsolescence in areas where the state of the art constantly changes. While an alternation between "exploitation" and "exploration" projects seems more relevant, the pauses between the projects are also crucial as they help us to enhance the capitalization of skills and the professionalization of individuals in their technical profession. A balance must also be found between continuity and renewal of the expertise areas [CHA 12]. Continuity not only makes it possible to give a greater depth to a field and develop valuable expertise for the individual and the company, but also promotes "closing up" in a specific field, which could bring problems of employability if the concerned field loses its scientific or strategic relevance. The frequent changes in subjects, on the contrary, allow for a greater versatility which is a guarantee of flexibility for the organization and for the individual, but it could endanger the acquisition of strong expertise within one particular field.

Another key issue raised by project organization is the selection of project managers. The complexity of this role has attracted much attention both within academic research and with the practitioners. Professional associations have proposed standard profiles, such as the *Association francophone de management de projet* (AFITEP) or the Project Management Institute (PMI), which develops standards and offers certifications. While there is not much debate on the competencies needed for a project manager, their relative importance can vary according to several factors: the nature of the project to be managed (exploratory or development), the project phase (the first phases of the project require creative strategist qualities, and subsequently, managerial needs appear) and the type of management (control focused on cost or profitability, see Chapter 6). Table 4.1 thus highlights the differences in the skills expected in a cost-control project (identified client, detailed specifications formulated by the latter) or in a profitability-control project (potential customers, technical specifications and deadlines are dependent on the internal decisions and are not prescribed by a clearly identified external sponsor).

Project manager's skills	**Cost- control projects**	**Profitability-control projects**
Mastery of the instrumental dimension of project management	+++	++
Mastery of the technical fields involved in the project	++	+++
Understanding of the project's specificities and commitment to its objectives	++	+++
Social competence	+++	+++

Table 4.1. *Project manager's skills by project type (according to Garel* et al. *[GAR 03b])*

The set of skills required for leading a project also depends on the power granted to the project manager, which could truly be a leadership role. As Midler [MID 93a, MID 93b] observed, the Heavyweight Project Manager designs the project, builds the teams, manages them and benefits from considerable delegation of power. The manager must therefore have strong skills and legitimacy with regard to the various project stakeholders since these roles are very demanding.

4.2.3.2. *Mobility of R&D professionals*

While project assignments could be considered as temporary mobility, R&D professionals could consider following their career trajectory both within and outside of R&D.

In this area, the trends are the same as those observed in the other functions. The HR function promotes a stronger mobility in all directions: organizational, geographic, functional, etc. This mobility is supported not only by the communication efforts, but also by the established procedures and mobility incentives. For example, at Rhodia, it was shown that, from the early 2000s, there would be no more internal promotion and, in order to apply for the managerial positions in R&D, the person should have passed through various mobilities. Sometimes, international experience was also required. Similarly, in the oil sector, passing through the operational position in the subsidiaries is a critical condition for career progression [LEL 14]. To facilitate these transitions, the HR function develops many tools. Some would support the individual proactivity, such as internal job databases, job mapping and possible career paths. Others are rather management tools, such as "people reviews" carried out by "career and mobility committees", made up of HR and line managers, who examine the situation of each R&D professional and point out possible evolutions for those who are considered to have stayed long enough in their current position.

In some companies, driven by the idea that people should not stay too long in R&D, there is strong pressure for R&D engineers to leave R&D and bring their ideas to other functions of the company. This is the pool model [GAS 07] in which R&D serves as a recruitment firm and training room for technical managers of the entire company. Within the framework of projects, the heads of other functions also more easily identify talented R&D engineers than before, and they seek to attract them with interesting proposals. Many R&D professionals are sensitive to these prospects of evolution. Indeed, career progress in "less uncertain" environments is often faster, remunerations are higher and extra-monetary compensation factors (responsibility, proximity to power and clients, pleasant workplaces, etc.) are more stimulating than in R&D.

Though these are strong trends, they raise a number of questions. Is the temporal pattern of mobility every three years (which is almost established as a norm) relevant to R&D, whereas its activities take place over longer periods? Since R&D's mission is to develop and establish a base of distinctive scientific and technical expertise that takes around 10 years to create, is it not risky to insist on classical rules of mobility for R&D professionals? Would it be better to keep a pool of highly skilled researchers and engineers in R&D who are able to invent radically new products from scientific breakthroughs? Here, we see the limits of the application of a standard human resource management model, which does not take R&D specificities into account. Some companies, on the contrary, try to make R&D an attractive career space in which professionals can move and evolve, without being required to leave the science and technology sphere.

4.2.4. *Individual assessment*

4.2.4.1. *Limitations of the standard model*

The assessment of R&D professionals is another HR practice, where the limitations of the standard model become evident.

The first limitation concerns the individualization of assessment. Indeed, HRM is moving towards the individualizing model as described by Pichault and Nizet [PIC 13], which is characterized in particular by the personalized assessment. R&D professionals are no exception. This orientation raises an initial series of questions in the R&D environment: "Does this include identifying the specific contributions of employees involved in essentially collective activities? Or is there a particularly subtle incentive mechanism that involves regularly recalling the spirit of collective work to people?" [MAR 03a, p. 6]. Such a reminder of the collective dimension of work seems essential, in view of the forms of work organization and performance drivers that characterize R&D today.

The second limitation in applying the standard R&D assessment model concerns the temporality of assessment: when should assessment be done? How often? R&D activities follow their own cycles, which are not always in line with the annual assessment logic. Some activities and projects may fall within a short-term horizon, while others may have much longer timeframes (for example, in the space and aeronautics fields where research programs may take many years to complete). Moreover, R&D activities, and in particular its research part, could hardly follow the annual budget cycle, as can be the case, for example, for sales representatives or accountants. The calendar year is not necessarily a relevant temporal pattern for these activities.

The third limitation in applying the standard R&D assessment model concerns the choice of the assessor: Who should assess R&D professionals? The standard response, which consists of saying "it's the direct hierarchy", is not always appropriate. As the process involves assessing highly developed and complex skills and the results of the advanced-level work, the hierarchy could not necessarily have sufficient expertise for the assessment. It is becoming even more difficult for managers, as the tendency is to enlarge departments and technical units by bringing together different technical areas. In addition, if the manager does not have a technical or scientific background but comes from technical marketing or industrialization, this could bring additional difficulties for the legitimacy of assessment.

The involvement of R&D professionals in projects also makes it complicated to rely solely on the judgment of the line manager. Indeed, they are not in a position to give their own assessment of the contribution of individuals in projects that they do not manage, particularly when these projects absorb a large part of the working time of researchers and engineers. Even if project managers forward their assessments, will it be possible for a line manager to give a fair judgment of contributions of an individual working on an external project, compared to the assessment of the individuals who remained directly under his/her supervision? Conversely, relying exclusively on project managers, who are engaged on a short-term horizon, is also not the ideal solution. They are inclined to strictly assess the specific qualities needed for project realization, which is normal, but these qualities may not include the specific indicators related to the professional criteria (which will be used in the specialized department), since the project manager does not necessarily have an appropriate reference framework of comparison. There is also a risk that the project manager could lose sight of employability and competence development issues of the assessed person that the business hierarchy could have in mind. Thus, line management still has a say in the matter, but it is henceforth no longer the only actor responsible for assessing the work of R&D professionals. Project managers, teammates, and peers within and outside of the company are also stakeholders in the assessment process.

A "good" assessor is therefore the one who is legitimate to judge the work of individuals, because of his/her skills and proximity to the activity carried out, which in the case of project organizations in R&D leads to a plurality of assessors to be put forward. Their understanding of the work performed and the skills mobilized also make them well positioned to propose competence development paths. Further to this, Galambaud [GAL 83] points out that the "good" assessor must also have sufficient influence in decision-making processes in order to be able to assure actions following the assessment (regarding the funding of training, premium distribution, assignments, work organization) and to put in place the resources required to achieve the objectives. Thus, the choice of the assessor must be subject to a more decisive question: assessing for what purpose? Leaving aside the bureaucratic idea that we assess "because we have to", the managerial response is that we assess to contribute to performance management in R&D, which raises the sensitive issue of assessing individual performance in this area.

4.2.4.2. *The sensitive issue of assessing individual performance*

The current trend is to focus such assessment on individual performance [GIL 17], whereas it could emphasize the way the work is done, the efforts engaged, behaviors, skills or development potential. R&D is not an exception to this trend, even though the limits of the assessment based solely on the estimation of results have been reported for this type of activity [MAR 03a].

There are many difficulties in assessing individual performance. How can a distinction be made between what relates to the individual and what relates to the situation or to other actors involved in the activity, knowing that the collective aspect of work in R&D is quite significant? Beyond that, since a plurality of actors can be involved in the judgment of the performance of R&D professionals, we are confronted with the diversity of the logics expressed by each of them (see section 4.1.1), as illustrated in Box 4.3. Chapter 6 shows the complexity of the very definition of R&D performance and its measurement. This complexity is found at all levels, whether it involves assessing the performance of a department, laboratory, program or an R&D engineer.

Three people meet to examine the results obtained by an R&D engineer of a large engineering company, and to discuss the engineer's performance. There is a technical expert (Researcher Emeritus), a business unit manager and an HR manager. Let us imagine the situation.

Everyone makes an assessment of the performance of this engineer and justifies such appraisal. We then observe the diversity of the judgment standards that everyone uses:

– The expert refers to the researcher's creativity, to the innovative solutions that he/she has brought to the technical problems raised and to the individual's reputation in his/her relevant scientific community.

– The manager puts forward the economic contributions of the assessed individual, the resources he/she consumed (time, money) and the financial potential of the contributions made.

– The HR manager reasons rather in terms of the gap between the objectives set at the beginning of the period and the results achieved, as stated in the annual performance assessment guide. He/she envisages recognition means and development measures (training, mobility, etc.) to be provided in accordance with the gap found.

Each of these points of view refers to a particular reference system in which diverse positive and negative elements could be in conflict.

Box 4.3. *The debate on the performance value of an R&D engineer (source: Bobadilla and Gilbert [BOB 15])*

4.2.5. *Remuneration*

Research work is known to bring intrinsic satisfaction (accomplishment, esteem, autonomy, skills development). However, on the one hand, not all R&D professionals devote most of their activity to research, and, on the other hand, the image of the disinterested researcher is largely mythical. Even if such a stereotype is internalized by young scientists when they enter this professional path, they do not withstand the material constraints that increase with age. The comparison that R&D professionals make with managers in other functions, who are often less qualified but better paid than them, generates frustrations and claims. As with other worker communities, remuneration is an important mobilizing factor (although not the only one) and it should not be neglected.

To meet these expectations, in addition to the fixed job remuneration, companies develop methods aimed at recognizing individual contributions. The introduction of variable performance incentives, within a set salary range for each job, is the most common action.

Another form of differentiated individual compensation is the remuneration for employee inventions. It is mandatory in France following a 1990 law and could take various forms according to the company strategy (see the example of Air Liquide in the box below).

Air Liquide Group is present in 80 countries with about 67,000 employees, and is the world leader in the area of gas, technology and services for industry and health. Innovation is one of the pillars of the group's strategy. It has more than 10,000 patents and files an average of 300 applications per year.

In 1997, Air Liquide launched its own inventors' recognition program, which applies to all of the group's inventors around the world. The program integrates the terms of the Collective Bargaining Agreement for Chemical Industries, but goes beyond such terms to integrate the international characteristics of the group. Supplementary compensation combines both a flat-rate bonus system and a variable part which is calculated on the basis of five criteria:

– The general research framework: this criterion evaluates the material conditions and the difficulties of creating the invention.

– The difficulties of practical development of the invention: for example, we know that the introduction on the market of a medicinal product (subject to clinical tests) is longer than that of a new electronic product. Average durations were established per activity sector, and completion earlier than the average period gives points.

– The scope of the patent: does the invention cover a whole range of new products or services? Or is it an improvement of a stage of the production process?

– The economic interest of the invention: this involves determining the benefits brought by the invention. Part of the assessment is based on the initial observed feedback and the other part is more prospective.

– The fifth criterion makes it possible to take into account other potential elements to be valued (productivity gains obtained, the contribution of the invention to the company's brand image, impacts on market shares, effects on the improvement of safety and the environment or even the detection of a talent).

Beyond remuneration, the program recognizes the work of inventors via an annual awards ceremony in a prestigious place. These events take place in the presence of the Group's management team in France, as well as in China, the USA and Germany, where Air Liquide operates.

Box 4.4. *Recognition of inventor employees at Air Liquide (source: Air Liquide, from Doyen and Fortune [DOY 16, p.19])*

Emphasis on individual reward remains dominant, although there are corporate managers who today could adopt Joly's [JOL 97, p. 82] formula: "The concept of the individual researcher who independently produces ingenious ideas is probably only a popular imagery".

Some companies have established collective bonuses, with a broad range of practices depending on whether the considered collective is the entire enterprise (with bonuses reproducing the logic of profit-sharing systems), R&D department or the project team. Thus, the rise of project organization has led to questioning of the relevance of an extreme individualization of remuneration. Indeed, the latter proves to be clearly inadequate, and even a source of perverse effects, given the significance of the collaborative nature of work, as well as the importance of the solidarity between project actors and their joint pursuit of project goals. Project organization also raises the issue of the temporality of remuneration: while it is logical to show recognition at the end of projects' key phases, the latter are not specifically connected to the periods set by the HR department for the award of bonuses. The establishment of team bonuses paid at the end of a successful phase of a project shows a certain adaptation of remuneration methods to R&D specificities. They remain, however, a rather rare practice.

Thus, regarding remuneration, as for other HRM practices, companies attempt to combine the consideration of R&D professionals' specificities, on the one hand, and general harmonization, on the other. This concern is at the origin of the "dual ladder" tool, which aims to propose an equal level of remuneration for technical experts and hierarchical managers. Since the dual ladder is considered primarily as a career management tool, we will present it in the next section.

4.2.6. *Careers*

4.2.6.1. *When career is not limited to hierarchical progression*

The concept of a career is a complex construct that has been the subject of numerous studies. Here, we adopt the definition proposed by Arthur *et al.* [ART 89, p. 8], according to which a career is an "evolving sequence of a person's work experience over time". This definition which is widely shared today by management scholars goes beyond the traditional view of a career, associated with hierarchical progression, to include any type of work experience that has meaning for a person.

This approach is particularly relevant for studying the careers of R&D staff for whom hierarchical progression is not always the dominant aspiration. Indeed, some give priority to specialization, preferring scientific and technical excellence to managerial responsibilities. Career aspirations vary according to the profiles, which can be quite diverse in R&D, even if there is no determinism. Thus, researchers, engineers and technicians, although they collaborate in their daily work, may have different values and expectations. The distinction introduced by Gouldner [GOU 57] between "locals" and "cosmopolitans" could also be found in R&D. Thus, alongside engineers who generally have "local" orientation, as they identify themselves more strongly with their company (by seeking to stay and progress within the latter, often within a classical and hierarchical sense of career), researchers, on the contrary, could be considered as "cosmopolitans" as they are generally more loyal and attached to their professional community, which they consider as a primary source of legitimacy and as a space where their careers are enacted[2] [LEL 14].

4.2.6.2. *From the accreditation to supervise research to the dual ladder*

The careers of industrial researchers have for a long time been modeled on those of academic researchers and, although this is less the case nowadays, there is still some porosity between these professional worlds. In France, some of the most "sought-after" profiles still prepare and present the

2 It is important to mention that the category of "technicians" and the specificities of their work are quite poorly studied in the literature. It is, however, of crucial importance that the company has a realistic vision of their expectations, in particular regarding career progression, in order to be able to respond to their aspirations.

HDR (accreditation to supervise research), which is defended in front of a panel composed mostly of academics. Thus, judgment on the value of work and individual competencies is externalized. It is fully carried out within the science and technology sphere, without the interference of managerial, economic, industrial or administrative logics.

Some companies have heard these expectations regarding expertise recognition from people who aspire to develop their skills and continue their R&D activities. The issues of compensating and reinforcing the loyalty of these experts are convergent with the challenges of strengthening and developing the expertise on which the company builds and which supports its innovation strategy.

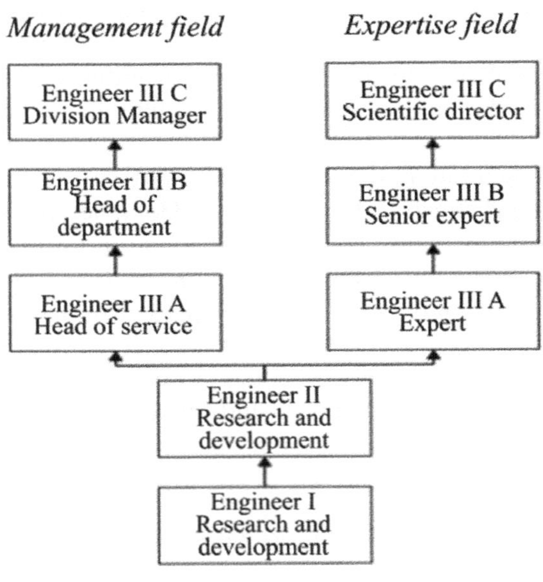

Figure 4.1. *A representation of the dual ladder scale*

Managerial tools such as the dual ladder career model (see Figure 4.1) attempt to value expertise and experts. Alongside the managerial path, a scientific and technical path is created that formalizes recognition and career progression for R&D professionals who distinguish themselves through their expertise. Such expertise is judged by an ad hoc committee, fully or partially composed of peers generally from within but sometimes also from outside of the company. In terms of remuneration, the salary levels are generally similar for an expert and a manager positioned at the same level on their respective career paths.

The dual ladder has gradually become the reference tool in experts' career management in industrial enterprises. It is however not devoid of difficulties and limitations. Thus, the expertise path that brings together the technical elite, highly selective, is ultimately undervalued, while the management path, though valued, is not very open to researchers [DUH 05]. The box below develops some of these limitations, and it should be noted that, despite the development of expert recognition practices (whereof the dual ladder is emblematic but not the only example), this remains a serious issue for HRM and an ever-existing challenge in R&D.

The dual ladder appeared in the 1950s in the USA in large R&D organizations that had difficulties attracting and retaining high-level professionals. The initial observation is as follows. Although the performance of scientific and technical activities requires the ability to rely on distinctive skills which are rare, non-substitutable, difficult to imitate and that take significant time to build, R&D professionals are, however, recognized neither for their skills nor for their contribution. Under these conditions, scientists and engineers seek to evolve towards managerial responsibilities that ensure them career progressions with better salary and status. Departures from R&D disrupt the dynamics of individual and collective skills development that is achieved over long time periods. Managerial responsibilities do not generally allow for individual expertise to be kept at a high level. This is particularly true in the research fields that are characterized by rapid obsolescence of knowledge and that, consequently, require permanent learning. The issue raised is therefore that of maintaining the motivation of those who, by choice or by default, remain in scientific and technical activities. To overcome these difficulties, the dual ladder model proposes a second career ladder, alongside the managerial path, which is based on the recognition of expertise. Professionals could progress on the expertise ladder through an ad hoc committee's assessment of the depth and breadth of their skills and scientific/technical reputation.

The first criticisms against the dual ladder model came shortly after its invention and they have been increasing ever since. Allen and Katz [ALL 86] identify excesses in the use of the dual ladder that lead to the degradation of the value of expert status in the eyes of the scientific and technical professionals for whom it was intended. Thus, instead of being used to distinguish and recognize scientific and technical excellence, the technical path is sometimes diverted to award "consolation prizes" to R&D scientists and engineers who are unable to progress on the managerial path because of their lack of interpersonal skills, leadership, relationships or internal supports. Using the dual ladder in this way helps to reinforce in the collective imagination the caricature of the "ivory tower researcher" type of expert.

Box 4.5. *The dual ladder: solution or problem? (source: Gastaldi and Gilbert [GAS 16])*

4.2.6.3. *What are the career paths for project contributors and managers?*

Regarding career paths, the life of project members, and especially project managers, is similar to that of expatriates. What happens after the project? Reassignment, indeed, raises a number of questions [ZAN 08]. Will the next project be as interesting? Is returning to the profession possible in good conditions? Will future missions (on a project or in the business) allow for the valuation of skills acquired during the previous experience?

Assignments to projects also raise questions about career progression prospects. Projects are not an avenue for the improvement of specialized skills, but rather for their expansion: they enhance generalist profiles and the acquisition of multidisciplinary skills. Projects value intrapreneurial approaches and group functionings. They thus produce profiles that are different from those valued in specialized professional communities. In general, we break from obvious and recognized progression paths in specialized departments. The return of a project member to a traditional path could thus be disturbing. On the one hand, this can be difficult for units that admit a person who is sometimes perceived as an unruly adventurer, and whose contributions and skills acquired in projects are poorly appreciated. On the other hand, this can also be difficult for the person concerned who is afraid of being caught up in a straitjacket.

Some companies have organized a specific career path for project managers, in addition to the hierarchical management career and, where it exists, the expertise career path. This is referred to as triple ladder. However, project work is intensive; it requires personal overinvestment and generates stress, in view of the emergency pressure and risks. Creating a specific career path for project managers thus raises the issue of the long-term sustainability of careers achieved solely in projects, with a chain of responsibilities for the project manager. In the case of companies having implemented a triple ladder, it therefore seems important to support the transitions (and not only put them forth in the rhetoric accompanying the deployment of this tool) from one path to another throughout the career trajectory.

4.2.7. *Competence management and training*

Competence management has been the subject of much attention in recent years. Though we can question its relevance in activities where knowledge is largely held by systems, conveyors of formalized processes, it is of paramount importance in R&D. In this function, learning takes a long time and is carried out in close connection with peers, within and outside the company. This situation raises the question of what the company can offer to highly qualified professionals, knowing that post-school learning is carried out according to an alchemy that goes well beyond management systems.

The HR function has thus tended to invest mostly in areas that are outside the scope of the scientific and technical field, to establish new training offerings in R&D around the development of behavioral skills, raising awareness on the company's strategic issues and reinforcing the capacity to cooperate with the other functions of the company. Thus, many large high-tech companies have set up training programs for their technical experts from different functions of the company and particularly R&D (see the example of Safran in the box below).

> Safran is a high-tech international group (aeronautics, space, defense, security) which has more than 66,000 employees, with a high proportion of engineers and managers (more than 40%), and which heavily invests in R&D (1.7 billion euros in 2016, representing 11% of turnover).
>
> In 2007, the group launched the nomination campaign of the group's experts, as well as a specific training program for this community. The training is provided once a year by Safran University (Safran group's training center) for recently promoted experts. The training consists of 4 modules lasting 3 days and spaced one month apart. They carry out a project in a team of 7 or 8 people, on subjects defined by the members of Safran's executive committee or by the directors of one of the group's companies. They are the sponsors of the projects that they have proposed and should be available to answer questions from the other experts.

Simultaneously, experts meet at conventions once every 3 years. They also participate in thematic days organized by expert groups on targeted topics. External professionals may also be invited to these meetings.

Box 4.6. *The training of technical experts of Safran Group (source: Safran information documents)*

Although the HR function brings added value by developing new extra-scientific training, it also tries to act upon the development of core competencies. Internal and external training is offered to enhance and maintain scientific and technical skills. The external training from private or public organizations is mobilized to this end. Equipment suppliers can also be crucial training actors as they provide essential knowledge on how to use highly specific equipment.

The development of digital technology is a major evolution and training services must adapt to it. This represents a promising solution for designing an offer adapted to R&D needs and, in particular, for personalizing training according to individual needs and aspirations.

HR managers should also strive to create the environment and conditions that encourage R&D professionals to provide internal training for their colleagues (their R&D peers or other professionals), in order to facilitate knowledge-sharing and mutual understanding between professions.

To summarize, the training of project members raises a number of specific issues. Strictly speaking, "being a project manager" is not a profession, even if some companies have created project career streams in order to recognize project managers and also to maintain the attractiveness of these crucial but complex roles which are not always "profitable" in terms of traditional careers. A professionalization of project managers is also observed. The Project Management Institute has established a reference system for training and certification of project managers that is required for major international projects. However, such training programs are still very rare and are often specialized around specific projects (computer science, construction, etc.) or project management methods. The action trainings could be very interesting as they offer project managers the opportunity to reflect on their own actions and management practices, supported by a coach, in order to improve their performance. This is done simultaneously with their regular work as a project manager. The difficulty of implementing such action training lies in the intense pace of projects, which does not promote reflexivity and provides limited time to participate in training courses which are, however, valuable learning and knowledge-sharing opportunities.

In order to adapt the training actions to this evolution, the HR managers should reflect on the specificities of R&D professionals and consequently adapt the methods of their training. This requires the training of managers to be quite open and proactive in the analysis of the learning process. It is important to consider and enact the "natural" resources of knowledge offered by the R&D ecosystem to facilitate and develop these learning processes. Thus, digital technology should not only be considered in terms of the development of e-learning tools ensured by companies or external providers, but also (and above all) as a means of intensifying and diversifying the modes of connecting R&D professionals with the external environment and epistemic communities including practices that enable valuable learning. The R&D professionals should therefore be given time for these learning opportunities and these practices should not be considered as "leisure" time. It should be the same for participation in external seminars and symposiums, as well as for reading academic and professional journals or monitoring patent databases (which are now easily accessible online but are not free of charge).

Focus must be redirected to the importance of the missions and projects to which R&D professionals are assigned (see also section 4.2.2.1), as these assignments would have crucial impact on development, maintenance, renewal, expansion or even exhaustion of the core competencies. Though these decisions do not generally fall under the responsibility of HRM professionals, it is nevertheless essential that they consider these issues in order to integrate the latter in staffing decisions. The partnerships with line and project managers on these issues are essential, and would help HR function to better respond to the challenges of R&D and to interact with technical and scientific staff in a more comprehensive way.

4.3. The new challenges of HRM in R&D

After characterizing current trends of HR management towards R&D activities, we will now try to question their relevance in view of the profound transformations that R&D is experiencing. It is thus possible to identify certain limits of HRM practices in R&D, which call for rethinking or even reinventing the management of R&D professionals. Section 4.2 has highlighted the challenges of HR practices, concerning in particular the recognition of R&D professionals and especially the experts. We focus here on three major challenges. The first is the contextualization of R&D staff management practices according to the diversity of contexts (section 4.3.1). The second concerns the adaptation of HRM to changes introduced by open innovation (section 4.3.2). Finally, we focus on the need to consider the collective aspect in HRM practices destined for R&D activities (section 4.3.3).

4.3.1. *Moving beyond an instrumental approach and adapting to the diversity of contexts*

There is enough literature to affirm that HRM must consider the real and precise contexts in which it operates to provide pragmatic responses that would be adapted to them [PIC 13]. Nonetheless, and despite all the criticisms that it has been subjected to, the instrumental model of HRM [BRA 93], which is based on the application of standardized programs of universal scope, still remains largely dominant today. Under this model, it would be up to HRM specialists to decide what is good for R&D professionals.

However, first of all, this requires critical reflection on how to adapt HRM management to the company's innovation strategy. This strategy as well as the modalities of the innovation process would vary significantly from one company to another according to the type of innovation (radical or incremental) and its pace. What are the individual and collective competencies needed to address these strategic issues? What are the means to develop these competencies (internal development or external acquisition)? For example, companies pursuing intensive innovation strategies, which require the ability to rely on scientific intrapreneurs in R&D [GAS 07], can hardly derive satisfaction from a dual ladder type of career management, as it will not respond to the aspirations of these individuals.

It is therefore necessary to adjust the HRM practices to the characteristics of the context that each company encounters. This requires HRM professionals to adopt design thinking, which will help them to go beyond the pre-existing tools. It also requires strengthening cooperation between HR and R&D managers, so that HRM professionals are able to design practices adapted to the particularities of innovation strategies, R&D organizations and the aspirations of R&D personnel.

4.3.2. *Renewing (or reinventing) HRM in open innovation models*

Open innovation practices displace some of the R&D activities outside of the firm, which requires the redefinition of roles within internal R&D (see Chapters 1 and 2). These changes raise a series of questions.

As the roles of internal actors will continue to evolve in a direction that is not fully obvious today, what will be the new challenges in terms of HRM of such roles? And what responses will HRM be able to provide? How will the recruitment, assessment and promotion criteria be adjusted to work transformations? How could the organization accompany the acquisition of new skills required by redefined roles? What could boost the motivation of professionals whose work content and environment are in permanent change?

In addition, HRM can no longer limit itself to the intra-organizational level. It should also build on the talent that can be found in the business ecosystem and that can participate greatly in the innovation process. What could the HRM actions be within the innovation spaces that go beyond the firm's boundaries (enabled by cooperations, partnerships, mobilities, etc.)? Some studies (namely Calamel *et al.* [CAL 11]) have focused on the consequences of the creation of competitive clusters on HR management in France. They point out the challenges of the development of new inter-organizational HRM practices that need to be addressed at the level of the territory: collective bonuses on collaborative projects, staff exchanges, common training programs, etc. Calamel *et al.* [CAL 11] also highlight the challenges that represent such innovative practices and their difficult adoption by traditional HRM. Much work still has to be done towards inventing an entirely new HRM, which is adapted to these atypical configurations that are becoming more frequent and can take a variety of forms.

Companies	Centralization/ decentralization of R&D	Structural changes	HRM practices
Company A (Food)	Greater importance of the central R&D structure, setting up of technological and scientific monitoring departments	From a functional structure to a matrix structure	Mobility of researchers within and between companies. Interaction between R&D and other functions Open dual ladder
Company B (Pharmacy)	Same	From matrix to network	Mobility of researchers within and between companies. Interaction between R&D and other functions
Company C (Specialty chemicals)	Same	From matrix to network	Attraction of talented university graduates Mobility of researchers within and between companies. Interaction between R&D and other functions
Company D (Aerospace)	Same	From matrix to network	Attraction of talented university graduates Mobility of researchers within and between companies

Table 4.2. *Consequences of the adoption of open innovation on the HRM practices of four multinational companies (adapted from Petroni* et al. *[PET 12, p. 170])*

Considering a range of open innovation practices, Petroni *et al.* [PET 12] observed how these practices transform the organizational structures of R&D and its staff's management methods (see Table 4.2). The open innovation strategy leads to the adoption of matrix or network structures, instead of conventional functional structures. It emphasizes the integration of external or multidisciplinary knowledge, rather than the enhancement of scientific sector-specific expertise. Thus, the companies will need the individuals with "T-profiles" (T men) who, beyond possessing specific skills in their scientific or technical area, would also have a good knowledge of related fields. In this regard, mobility management, organized both internally (differentiated career paths) and externally (temporary assignments with academic laboratories or jobs within subsidiaries), is a key practice.

In the context of open innovation, managing expertise cannot be limited to managing a body of experts and internal expertise. The challenge also involves coordinating internal and external resources, by going beyond the usual HRM and expert management organizational framework (see Chapter 3 for avenues on the subject). The objective of the company is to ensure, through the consideration of the environment and the major developments in scientific and strategic spheres, the flexibility of the expertise on which to rely on.

Looking at the multifaceted nature of open innovation practices and the change they bring for management structures and methods, we can see the enormity of work to be done!

4.3.3. *Going beyond individualized HRM by integrating the collective dimension*

Finally, we will focus on the importance of group dynamics in the creation and dissemination of knowledge. Whether it is a project-based structure [GAR 03b], or the maintenance and development of expertise within specialized professions [LEL 14], the collective dimension seems to be paramount in terms of both competence development and expertise recognition. The challenge is thus to find the right balance between the collective and individual dimensions. However, there is a tendency to neglect collective competence in current HRM systems. We thus propose to overpass the individualization trend, by implementing practices and management tools aimed at promoting and supporting collective expertise and collective competencies rather than enhancing individualizing practices. The following chapter proposes the detailed analysis of such practices and tools.

Let us add in consonance with the previous section that this call for a more collective and shared HRM should be considered within the current context of open innovation. For HRM, this would also involve projecting itself beyond the strict boundaries of the organization. What a challenging and also passionate agenda for HRM for the years to come!

4.4. Conclusion

Although HRM practices are now clearly more present in R&D, which could be seen as a consequence of an increased willingness to rationalize and manage these activities, some of the practices present certain challenges, and sometimes even problems. While R&D is constantly changing and striving to adapt to current transformations (chrono-competition, ambidexterity, globalization, digital revolution, open innovation, etc.), HRM must constantly question the relevance of its methods destined to the specific communities of R&D professionals. The HR function has to overcome major challenges in order to valuably support the ongoing R&D changes, some of which are still in their infancy and need thorough interpretation and analysis. The avenues for rebuilding a new HRM mindset could first emerge from the consideration of the real conditions of the R&D activity, the diversity of contexts and the current challenges of intensive innovation, second, from the incorporation of R&D activities in open and complex ecosystems and, finally, from the collective dimension of work and performance in R&D. This reinvention of HRM requires a strengthening of relationships and cooperation between HRM actors and R&D stakeholders.

5

Collective Expertise: Forms and Methods of Management

While managerial practices are shifting toward individualization, the importance of collective action is becoming more than ever a strategic issue for the company. Indeed, for R&D activities, which often target breakthrough innovations, the ability of people to collectively collaborate and create becomes an undeniable asset. Such collective action often requires the coordination of varied and complementary expertise, which makes it possible to deal with the complexity of the projects and missions in which R&D professionals are involved. It is therefore crucial for technology companies to be able to identify strategic processes that require collective expertise and to support actions in which specialists and experts collectively commit themselves.

In this chapter, we focus on the different forms that collective expertise can take within an organization and the different management mechanisms that can support them. In the first section, we highlight the different aspects of organizational expertise and underline the importance of its collective forms, particularly in view of increasing individualizing management practices. In the second section, we present a detailed analysis of different forms of structuring collective expertise in an R&D context, by distinguishing between interdisciplinary and monodisciplinary communities of expertise. For each of these communities, the prospects and challenges are pointed out, including the management systems, thus allowing better management of the collective action of R&D professionals. We conclude by highlighting the importance of coordinating the different forms of collective expertise in order to reinforce a learning dynamic and ensure the maintenance and development of critical expertise in the future.

5.1. Collective expertise in R&D

In order to remain competitive in a global knowledge-based economy, industrial companies seek to distinguish themselves by using rare, advanced and well-managed expertise. The expertise path thus becomes an alternative to the management path and the expert status gradually increases in visibility and legitimacy. However, the concept of organizational expertise still remains underexplored in management sciences and each company is inspired by its own vision of what it considers as real expertise, who its actors are and at what level they should act to develop and maintain such expertise. In this chapter, we attempt to clarify the concept of organizational expertise by focusing on its collective dimension.

5.1.1. *The dual facet of expertise: individual attribute and collective process*

In academic publications, expertise has been analyzed through two dimensions: as individual attribute and as collective process. Cognitive psychology develops this first aspect of expertise [ERI 06] by focusing in particular on its acquisition and development patterns. The mental models of experts and their ability to process and organize a large amount of information that is acquired through deliberate practice over long periods, are regarded as key factors that explain experts' "talent". Experts are thus perceived as brilliant individuals whose abilities and performance are far superior to those of laypeople.

Other trends in academic literature and especially studies in social psychology [MIE 01] and education science [EDW 10] oppose this strictly cognitivist vision of experts and propose to situate the latter in their context. Expertise thus becomes a relational and collective object. Indeed, the "expert" is seen as a social function, which is carried out through the interaction of the expert with non-experts [MIE 01]. Edwards [EDW 10] reinforced this relational turn by criticizing the view that defines experts as heroes who are rewarded for their autonomous and independent work. She showed that professional practice emanating from the collaboration of multiple experts is more advantageous than autonomous action. Similarly, work on transactional memory [WEG 86] supports the idea that the creation of group memory is conditioned by the transmission of knowledge among group members in their respective fields.

Both dimensions (individual and collective) are crucial for understanding who is "expert" in an organizational context [LEL 11]. Individual characteristics, such as scientific excellence, openness and the ability to provide technological leadership over a specific field are indispensable assets that allow an individual to acquire an expert status. However, this status probably remains illusionary if experts do not put their expertise at the service of many organizational actors. The enactment of experts' knowledge implies collaboration with those who are in need of expertise as well as with peers both inside and outside of the company. The company must therefore provide a conducive environment that helps in enhancing the knowledge of individuals as well as fostering opportunities for experts' collaborative action in order to make their role "real" within the organization.

5.1.2. *Collective expertise and its current status*

Individualization of management practices, especially those of HRM, is a global trend, and the R&D function is no exception to this [LEL 18]. Assessment, remuneration, career management and training policies tend to adapt to the specific characteristics of a particular employee: assessing and rewarding individual rather than group performance, as well as prioritizing policies that strengthen competitiveness rather than collaboration. Such practices require employees to have strong negotiation and leadership skills. By putting individual needs at the forefront, companies give more autonomy to individuals, but at the same time delegate to them the responsibility of managing their own career and professional development.

This individualization trend may prove difficult when it concerns managing expertise. At present, management practices focus more on the issue of recognition of experts, particularly through assigning expert status while implementing a dual-ladder system (see Chapter 4). As such, companies hope that they can make the expertise path more attractive and present it as a real alternative to the managerial career. However, the prevalence of individualizing management tools, without coordinating them with practices aimed at the collective, can hamper the development of expertise and undermine strategic processes such as learning, transmission of knowledge or even decision-making. Studies show that the success of these processes largely depends on the quality of interactions between professionals [LEL 18].

Although a call for new, more collaborative and transversal organizations can be found in several managerial discourses nowadays, concrete practices do not usually follow. Initiating this change is particularly difficult as there is no clear understanding of how collective expertise emerges and what forms it could take in organizational contexts, which, in turn, makes it difficult to propose appropriate management policies. It thus seems important to define the different forms of collective expertise, explore their possible interdependence and specify the management mechanisms that will help to enhance and support them.

5.2. Two forms of structuring: "horizontal" and "vertical"

Collective expertise can take many forms within the organization according to the purpose and methods of collaborations between R&D professionals. Such collaborations can serve as the basis for the creation of different types of expertise communities with different objectives and temporalities. Depending on the modes of interaction, we can distinguish between two forms of structuring of expertise and thus two types of expertise communities. On the one hand, "horizontal" structuring of expertise means that professionals from different disciplines come together to address a particular problem or seek innovative solutions around a theme requiring a multidisciplinary approach. On the other hand, "vertical" structuring of expertise includes the collaboration of people from the same discipline who share common knowledge and have a similar set of values and norms. The challenges and collaborations are different between these two types of communities.

5.2.1. *Horizontal structuring: interdisciplinary communities of expertise*

5.2.1.1. *Defragmentation challenges*

The development of interdisciplinary and transversal expertise raises multiple challenges for a company. With the growing complexity of technologies, processes and services, companies are called upon to increasingly manage ambitious projects, which require combining diverse expertise in order to address complex issues. Indeed, the collective action of professionals within collaborative interdisciplinary communities also makes it possible to consider new avenues of exploration and thus encourages innovation. However, the success of transversal work depends on the organization's ability to provide resources and autonomy to professionals in order to initiate a fruitful exchange of ideas and contributions.

R&D professionals may, depending on the needs of the company, be required to work in several types of communities of expertise. These communities vary depending on the duration and frequency of professionals' intervention (short-, medium- or long-term intervention, occasional or permanent participation) as well as according to the types of knowledge that professionals implement during this intervention (institutionalized or emerging knowledge). First, engineers, scientists and technicians in R&D departments may be invited to work on the resolution of urgent problems or, in other words, to carry out "troubleshooting" activities. They also participate in R&D projects with rather medium- and long-term temporalities intended to develop new products or services based on research data. Alongside these temporal communities, R&D professionals with the recognized expertise level may be encouraged to contribute to permanent communities of expertise, for example, a "college of experts", aimed at bringing the company's experts together to discuss multidisciplinary issues.

5.2.1.2. *Troubleshooting teams*

The term troubleshooting refers to interventions which aim to identify the source of a problem. Troubleshooting is a key activity for many industrial companies. The ability to provide emergency response to solve complex problems becomes a key issue, which makes it possible to avoid malfunctions and thus maintain the performance of facilities, services and products. The quality of troubleshooting has a direct impact on companies' reputation, because often the technological issues could not only influence the internal functioning of the company, but may also have significant societal impact (consumers' health, safety of facilities, etc.).

To perform troubleshooting activities, the company can rely on high-level in-house expertise. Specialists and experts are called upon to identify the causes of problems and explore avenues that might offer solutions. Interventions are mainly short term, but some types of problem may require long-term interventions, for example when a shutdown of a technical unit is required.

Depending on the internal organization of each company, troubleshooting activities can be more or less formalized. In some companies, there are specialized units that receive requests for interventions and try to find appropriate expertise within the company. For this, they can rely on internal expertise databases. In other cases, the request for expertise can be done directly and in a less formalized manner by a person who comes across a malfunction.

The troubleshooting process passes through many stages. It begins with a diagnostic phase to identify the causes of a detected anomaly. Depending on the problem's complexity, several scenarios could be considered in order to find an optimal solution to the problem. If the chosen scenario requires direct expert intervention, it will work in collaboration with the local teams that requested the expertise, trying collaboratively to solve the issue, restore the system (technology, service or product) to its original state or even improve its original characteristics. Once the problem is resolved, follow-up is required to make sure that there are no new malfunctions. Finally, a reporting phase regarding accidents is most frequently organized in order to capitalize on the problems encountered and foster organizational learning.

> The complexity of the oil industry's facilities and processes requires Total, a major actor in the sector, to have the ability to anticipate problems and provide emergency response to technical challenges. To that end, a team of specialists and experts was formed to ensure troubleshooting for all of Total's industrial sites. This team has about 10 people who can intervene on the varied issues, such as damaged equipment, corrosion or other types of problem. Each team member is a referent for one or more industrial sites, but may also be called upon to act as an expert according to their specialty. Depending on the nature of the problem encountered, the referent identifies a specialist or an expert who can further act autonomously to the resolution of the problem. However, for more complex problems, intervention by a group of experts may be necessary. This may in particular concern technical equipment, which is essential for oil platform activities, such as furnaces. In fact, Total's facilities have many types of furnace of which the maintenance requires knowledge in operations, metallurgy and many other technical fields. If a malfunction is found, then the troubleshooting team referent will bring together specialists and experts in these areas to find the cause of the problem and collectively develop a solution. Various tests are then carried out to check the proper functioning of the facility and the absence of new anomalies. Finally, the type of problem and the solutions considered and applied will be formalized in an internal document called the Troubleshooting Guide. This will capitalize on the knowledge acquired and comprise a notebook of good practices.

The capacity to manage the divergent interpretations and opinions related to the specificities of each expert's respective field remains, however, to be one of the most critical issues in troubleshooting. Interpersonal skills thus become an undeniable asset required to find compromises and ensure smooth conduct of troubleshooting.

Box 5.1. *At the service of emergencies: "troubleshooting" activities at Total. Source: interview with François Menard, Total Expert*

As demonstrated in the case of Total, the use of a single expert may be insufficient and the intervention of a group of experts may be necessary to match various disciplines and allow better analysis of different aspects of a phenomenon. Indeed, research shows that transversality and collaborative work initiate creative moments that are required for problem-solving, which further helps to generate new interpretations and widen the range of possible solutions. The initiation of such creative moments is, however, subject to certain conditions [HAR 06b]. Collective action by experts is only possible if the problem is already identified or if there has been a request for expertise. Depending on the organizational environment and the company's practices, such help-seeking activity can be formalized or remain informal. This will determine the form of collective action. Second, everyone's willingness to dedicate their time and collaborate effectively will be crucial for the success of help-giving activities. Finally, the ability to review initial issues, by adopting new interpretations that arise in reflective reframing, makes it possible to consider often unexpected and original new solutions.

The success of these three processes, which is crucial for the initiation of creative moments, is conditioned by the collective action-friendly environment. The ability to request and provide assistance, review initial issues and collectively find new interpretations must be part of the corporate culture, valued on a daily basis and rewarded through the processes of assessment, compensation and career management.

Moreover, some logistical efforts seem necessary. In order to be able to provide quick response in emergency situations, companies must maintain and update expertise databases and encourage exchanges between potential expertise seekers and experts. The forming of troubleshooting teams also requires particular attention to both ensure the complementarity of the knowledge to be applied and anticipate the quality of interactions between team members.

5.2.1.3. *R&D projects and transversal expertise*

As discussed in Chapter 2, in order to meet the requirement of competitiveness, a large number of companies have opted for "project-based" organization, structuring the latter on the most promising strategic axes. The success of these projects largely depends on the quality of the interdisciplinary work between R&D professionals who put their varied expertise at the service of the creation of products and services. Unlike troubleshooting activities, R&D projects often require a longer commitment. However, this period can significantly vary depending on the industries and life cycles of the products and services involved.

There are several types of R&D projects. Differentiation can be made according to the level of change involved: exploratory projects aiming at radical innovation and projects based on the improvement of one or more product and technology characteristics that enhance incremental innovation. Some authors also distinguish "hybrid" projects, which bring together the two types of innovation targeted by the same project [CHA 15].

The success of interdisciplinary work within R&D projects is subject to several conditions: clear objectives that are shared by all, complementarity of profiles within a team, effective communication, availability of necessary resources, valuation and recognition of projects members' contributions.

However, even when these conditions are met, project teams may encounter many obstacles, because an interdisciplinary work within research projects is far from being a trivial task [EDM 09].

First, the complexity of the missions within R&D projects could give rise to ambiguity and uncertainty, related in particular to technological aspects, since the technology may not be in line with the identified needs. Changing market dynamics may also be a factor of complexity, because some disruptions could compromise the project's objectives and sometimes jeopardize said project.

Another difficulty has to do with the functional and disciplinary diversity of the project teams. Although beneficial for innovation, diversity could be a source of disagreement. Differences in views and opinions as well as discrepancy in analytical patterns and interpretative models are quite frequent. It is therefore essential to create conditions that prioritize the psychological safety of team members in order to allow them to freely exchange ideas, be innovative, commit errors and, above all, learn.

Another challenge is the temporal nature of these projects. R&D professionals are required to participate in these projects according to their area of expertise. Some will be committed for the entire project period, while others will occasionally intervene in certain phases or on particular aspects. This temporality necessarily implies an adaptation period for a new team in order for them to get to know each other, find the best communication strategies, apprehend everyone's expertise, share ideas and find the required consensus.

The fluidity of the boundaries of teams working on R&D projects does not simplify the collaboration challenge. The high turnover among team members can compromise learning and capitalization ability. Moreover, the advent of digital technologies, which makes it possible to access geographically dispersed expertise, also presents challenges related in particular to the difficulty of teams in communicating and sharing information without a physical presence.

Finally, the commitment and contribution of individuals to R&D projects must be valued and rewarded at the organizational level. This often implies the need for rethinking the reward system in order to integrate the criteria that value collective contribution.

The scientific and technological expertise of L'Oréal is well known. In order to recognize this expertise and give visibility to expert work, the company decided to create an expert path alongside the managerial one. Although scientific and technological excellence is a prerequisite for the attribution of expert status, the experts' ability to put their expertise at the service of L'Oréal's strategic activities was also considered essential.

Historically, L'Oréal's research and innovation activities are organized around three main areas, namely advanced research, applied research and development. Each area has its unique expertise, which enabled L'Oréal to develop and patent innovative molecules, one step ahead of its competitors. Currently, L'Oréal is preparing for a new strategic milestone by strengthening collaboration between different expertise areas and combining the various fields of expertise. Thus, the transversal projects that are structured around strategic themes for the group aim at making better use of available resources and allowing a new dialog between the various fields of expertise. Experts' contribution to these projects cannot be underestimated. Their scientific and technological excellence stands as a guarantee for the proper functioning of these projects, allowing the validation of strategic choices or providing specialist knowledge to address key issues. However, beyond the ability to offer technical solutions to identified issues, experts are particularly expected to broaden the scope of reflection by interpreting subjects and treating them from a different perspective. According to Gilles Spenlehauer, Director of Advanced Research: "The expert's role is not to bring technological solution, but rather to explore new possibilities".

To ensure the presence of this innovative spirit, the expert must be able to identify opportunities and areas of concern both internally and externally. Thus, experts' contribution to transversal projects is structured around the following two aspects:

– identifying the ecosystems of knowledge and potentially interesting concepts for L'Oréal and

– analyzing opportunities for collaboration with the best university teams who have the necessary expertise.

However, the success of such interdisciplinary projects is conditioned by a strong expertise base that is deeply rooted in the organization. There is a risk that the experts would be dispersed too much between the different projects, which will give them little time to maintain their expertise at the required level of excellence and to pass on their knowledge to others.

Box 5.2. *The contribution of experts to major cross-cutting projects. Source: interviews with Gilles Spenlehauer, Director of Advanced Research at L'Oréal, and Jacques Playe, Director of Information Systems and Digital Development at L'Oréal*

Beyond the challenges of launching innovative products and services that help the company to develop its competitive advantage and position itself as a market leader, it is worth mentioning another aspect related to collaborative work. Participation in research projects also offers R&D professionals the privileged moments of learning. Interdisciplinarity, the diversity of situations to manage and the pursuit of common objectives allow them to deepen and broaden their field of expertise.

Researchers distinguish two types of learning emanating from R&D projects, namely intra-project and inter-project learning [CHR 15]. Organizations often seek to formalize learning within a single project, by putting in place assessments, capitalization tools and sharing sessions on acquired knowledge or skills. However, very little reflection is conducted on the ability to learn and develop expertise through a series of projects. This lack of a systemic approach generates a very partial and static vision of learning and the creation of expertise, since this does not enhance temporal and dynamic analysis. However, this task is far from obvious. Such an approach requires not only an efficient infrastructure and an adequate communication system for the monitoring and coordination of inter-project learning, but also the creation of new organizational roles and new types of interactions. The role of R&D managers in particular is reconsidered not only to ensure a necessary supervision of the matching of needs to resources, but also to jointly coordinate, with project managers, all professionals' learning paths through their successive participation in projects. This ability to integrate individual learning into collective expertise should allow, in a temporal perspective, a dynamic and coordinated management of knowledge and career paths to be ensured.

5.2.1.4. *Colleges of experts*

Specific types of community have emerged within industrial organizations. These are communities made up of experts from various fields who seek to use their knowledge to contribute to the development of new strategic development paths for the company. These communities, which could exist under different names ("colleges of experts", "communities of the future" and "communities of experts"), are positioning themselves as levers of innovation and indispensable strategic decision support tools.

The creation of colleges of experts often follows the implementation of the dual-ladder system in an organization (see Chapter 4). It aims at enabling experts to contribute collectively toward the development of strategic areas by relying on exchanges between different fields of expertise.

These communities are made up of working groups that could have a varied structure and life cycle. Some are structured according to targeted contributions, by distinguishing, for example, subjects relating to radical innovation on the one hand and incremental innovation on the other, as is the case at STMicroelectronics [CAB 17]. Others are instead organized by strategic themes that require multidisciplinary expertise, as is the case at Orange (see Box 5.3). Most communities have a governance structure (the members of which are usually elected from among the experts) that ensures overall functioning and establishes a link between the different working groups. These communities have varied activities: benchmarking, identifying prospective avenues of exploration, analyzing new technologies, training, knowledge management and so on.

In 2010, Orange created the expertise path and opted for limited-period expert status by emphasizing the importance of not only maintaining expertise, but also the contribution of experts toward the strategic development of the company. Such contribution was considered to be an essential condition for renewal (or not) of status after three years.

In order to foster this contribution and create the conditions enabling experts to work collectively on the company's key issues, Orange initiated the "Orange Expert Communities" forums for interdisciplinary thinking, which bring together experts from diverse areas of expertise. The following six communities gradually emerged within Orange: *Réseaux de demain* (Networks of Tomorrow), *Solutions pour Services de contenu* (Solutions for Content Services), *Sécurité, Énergie et environnement* (Safety, Energy and Environment), *Exploitation de réseaux* (Network Operation) and *Software et Services de communication* (Software and Communication Services). Each community is organized into working groups that meet on a temporary basis and aim at identifying strategic development prospects for the company. These groups are required to produce targeted studies or analyses, often in conjunction with other people inside or outside of the company who may be invited to participate in their work. The collaborations between groups are encouraged in order to address all of the dimensions of the issues under exploration.

According to Prosper Chemouil, Lead Expert at Orange, the objective of the communities is to anticipate the future critical developments for Orange and work on the possible avenues of action: "The objective is to better anticipate the changes witnessed by the telecommunications sector and to analyze how Orange should position itself and act. Orange Expert Communities exist to provide possible solutions to decision-makers to act strategically."

Orange Expert Communities cover varied subjects including digitalization, sustainable development, the circular economy, connectivity of objects and artificial intelligence, which have not spared the telecom sector. Several working groups are regularly formed within the Orange Expert Communities to investigate the potential impact of these trends on the company and recommend possible actions.

After conducting their research, the working groups share their results through seminars or reports, thereby contributing to the decision-making process. If the subject represents an important issue for the company, then it could be transmitted for an in-depth analysis within the R&D department.

Even if the choice to join working groups is done on a voluntary basis (according to the sensitivities and expertise specificity of each one), each Orange expert commits to dedicating 10% of their time to the expertise community. Such contribution is an important assessment criterion for expert status renewal. Within these groups, experts are encouraged to work on interdisciplinary topics, which require going beyond the scope of their area of expertise and exploring issues at the forefront of several fields. Thus, according to Prosper Chemouil, the ability of experts to go beyond the limits of their field and work transversally is of primary importance: "[The goal] for Orange, is to remove experts from their everyday environment. Someone who works on a subject must have the free mind to get interested in another area. […] In a community, experts are not only required to contribute within their field, because they are already devoting most of their 'usual' time to such assignments; they are also expected to be interested in what others are doing."

This multidisciplinary work enables experts to question their field of knowledge, test their hypotheses and assumptions in fields other than their own and thus broaden their competencies and their vision. This also allows Orange to integrate new knowledge, identify promising avenues of action, provide technological leadership in future challenges and also identify areas that require increased competence development, especially regarding new technologies or approaches.

Box 5.3. *Orange expert communities. Source: interview with Prosper Chemouil, Lead Expert at Orange*

5.2.1.5. *Challenges of interdisciplinary work: roles, interactions and knowledge creation*

Collective action of experts carried out in an interdisciplinary manner has significant advantages for individuals as well as the organization. In fact, transversality is often considered by researchers and practitioners as a primary means of initiating innovation. The crossing of different disciplinary fields helps to broaden individual knowledge and avoid fragmentation. It is also an effective way of knowing and recognizing each others' expertise. Moreover, transversal work could be very helpful for decision-making processes: a collective opinion of experts will have more legitimacy than an individual opinion and would thus enable experts to further contribute to the company's strategic orientations [LEL 18]. Finally, this transversality also contributes to strengthening the collaborative culture that generates a better circulation of ideas and expertise within the organization.

At the same time, such collective and collaborative work faces several challenges at relational, organizational and strategic levels.

Relational aspects play a significant role when it comes to collaboration between people who have in-depth, well-established and deeply rooted expertise in a specific area of knowledge. Moving out of one's disciplinary field to consider related fields requires a certain openness and willingness to confront and accept other interpretations and analyses. Under such conditions, collaboration may only be beneficial if there is good understanding and mutual respect between participants. It also requires the organization to ensure the complementarity of expertise and equally take into account the interactional capacities of the people who will be working together.

Whether it involves troubleshooting activities, working within R&D projects or participation in the college of experts, the contribution of R&D professionals to these different communities must be supported and valued through different organizational processes. Such contribution must specifically be taken into account during annual performance assessments and may become one of the major criteria for the development of a professional career. The ability to carry out interdisciplinary work requires specific skills, and the company must strive to foster their development. Beyond efforts to institutionalize a collaborative culture, special training courses could be provided to both raise people's awareness with regard to the issues and forms of collaborative work and develop their skills, particularly in the coordination of cross-functional teams. Despite the autonomy of R&D professionals and their strong preference for self-control over their resources, a certain degree of leadership may be needed in order to coordinate and implement collective action.

Finally, from a strategic perspective, it is essential that the results of the collaborative work of R&D professionals are taken into account by decision-makers and considered as a valuable support for defining future orientations of the company. Otherwise, the motivation of these highly skilled professionals to engage in transversal work could be compromised.

5.2.2. *Vertical structuring: monodisciplinary communities of expertise*

5.2.2.1. *Learning and unlearning*

Although currently organizations are more and more interested in transversality, the challenges of maintaining and developing strong expertise around critical areas are however no less important. The challenge is twofold. This involves not only learning new skills but also avoiding the loss of well-mastered knowledge. The departure of an expert or restructuring of a service are some of the factors that are likely to destabilize organizational memory around a field of expertise and also compromise its legitimacy both inside and outside of the company (with regard to its customers and partners). The phenomenon of organizational forgetting is increasingly attracting researchers' attention, and multiple studies are being carried out to understand the causes and consequences of this process [MAR 03b]. Forgetting is not always detrimental for the organization. In some cases, this can increase organizational performance, especially if it concerns forgetting inefficient practices or processes to relearn new ones. This is intentional forgetting, which can be desired and sought by the company especially in the context of change management. However, in most cases, organizations face unintentional forgetting that is likely to compromise their ability to master a technology or find expertise.

Despite investments in knowledge management, organizational forgetting is becoming more than ever a concern for technological companies. Indeed, some strategic choices could be disruptive for organizational memory. On the one hand, to reduce costs and improve flexibility, companies tend to outsource certain types of expertise. Such outsourcing, which originally concerned "support" functions such as information systems and some sub-fields of HRM (e.g. payroll services), then spread to certain "core business" expertise that could provide a competitive advantage to the organization. On the other hand, as shown above, companies are increasingly developing "project-based" organization to foster multidisciplinary work [EDM 09] in order to reinforce innovation capacity.

Both strategies can have a significant impact on the company's ability to manage expertise over the long term. Several examples have shown that the total or partial outsourcing of specialized resources can cause the loss of in-house expertise because of the inability to address certain organization-specific problems due to the absence of critical knowledge and competencies. The organizational memory is thus compromised, which could have long-standing consequences for the use and development of expertise.

Similarly, too much transversality (particularly through "project-based" structuring) could possibly jeopardize the ability to develop an in-depth expertise that generally takes about 10 years to build. Thus, the development of a solid "intradisciplinary" expertise becomes a critical issue for the company and a necessary condition for generating efficient transversal work.

Consequently, the role of organizational communities where specialized expertise can be accumulated, maintained and passed on to future generations should not be overlooked. Monodisciplinary communities of expertise thus become a favored space for the development of strong in-house expertise.

5.2.2.2. *Monodisciplinary professional communities: positioning and status*

According to Barley and Tolbert [BAR 91], companies need to have permanent access to specialized expertise internally. In technological and industrial companies, this expertise can accumulate within monodisciplinary communities of expertise [EYH 03, LEL 14] made up of members who share the same body of knowledge and work together to respond to expertise needs. For these professionals, communities provide a favored environment for learning, socialization and professional evolution. For the company, the dynamics of such communities ensure the quality of expertise and guarantee the maintenance of organizational memory.

Monodisciplinary communities can have various forms and statuses. Some communities are organized into specialized departments and have a formal status and visibility in the organizational charts. But corporate recognition is not a prerequisite for the existence of a community. These communities can bring together professionals who are not geographically and/or functionally connected, but share the same profession, the same identity and common professional values, and are required to work together on specialized expertise needs.

The positioning of such communities, as well as their evolution within the company, depends on the overall organizational strategy. The strategic orientations of the company, its competitive environment and technological choices could both initiate the emergence of new communities and cause the disappearance of some others. This in turn can have a significant impact on the career paths of R&D professionals. The presence of a strong professional community, deeply rooted in an organizational structure, will be a considerable asset for those wishing to choose a specialization path. This makes it possible not only to benefit from peer exchanges and to learn new developments in the field, but also, above all, to minimize the risks of specialization by acquiring recognized and valued expertise. On the contrary, the absence of the established community can pose some risks because knowledge is not yet formalized and their strategic character is not certain. Indeed, it involves exploring the unknown with uncertain results, which can possibly endanger the career development of those who engage in such exploration. At the same time, such uncertainty also offers the opportunity to be at the origin of the development of a new area of expertise, which could be a promising starting point for becoming an expert in this newly created field.

The monodisciplinary professional community, due to its strategic nature for the maintenance and development of key knowledge, becomes an important management object, though little known by managers. In order to understand how such a community works, it seems relevant to focus on the knowledge dynamics as well as the roles and interactions of specialists and experts within these communities.

5.2.2.3. *A new management object*

The benefits of considering the monodisciplinary professional community as a new management object lies in the need to manage two processes simultaneously: managing specialized knowledge dynamics around a specific area and managing individuals who are the holders of this knowledge, their learning dynamics and their career paths.

Analyzing roles and relationships within such a community brings a valuable contribution that allows us to define the avenues for managerial actions. The concept of role is considered as an essential element in understanding organizational behaviors [ALV 05, KAT 66]. Role refers to a set of activities and behaviors expected of a person [KAT 66] within an organization. If these expectations are not clear or contradictory to the person's perception, role ambiguity or role conflict may arise. The consequences can be harmful both for individuals (e.g. work-related stress caused by uncertainty) and for the organization (loss of key knowledge, ineffectiveness of professional relationships, declined performance, etc.). It is therefore important to clarify organizational roles and to ensure that there is no erroneous interpretation that is likely to generate ambiguity or conflict. But beyond the clarification of roles, the challenge also lies in the capacity to enact these roles in order to make them a real resource for individual and collective progress.

Within the monodisciplinary professional community, each person performs a specific role. The strong interdependence of each role and their complementarity make it possible to better meet expertise needs. Thus, when one of the organizational actors requests expertise, this will initiate the expertise relationship, in which all of the roles will be activated to provide the necessary collective expertise. The key to understanding the functioning of the community lies in the ability to provide a detailed analysis of all its roles by differentiating them with each other. It helps in understanding the possible paths of progression between these roles over time. Finally, such analysis makes it possible to initiate a management action to manage the roles and missions of a community's members, the knowledge on which these roles are based and paths of progression between these roles. As a result, the career choices available to professionals become visible.

Research, carried out at the *Centre de Gestion Scientifique de l'École des Mines de Paris* (Center for Scientific Management at the Ecole des Mines, Paris) has helped to refine a management tool that facilitates the visibility of a role structure and the pathways of a monodisciplinary professional community, as well as to influence the dynamics of its evolution [EYH 03, LEL 14].

The tool proposes a mapping of roles along two axes, namely technical expertise and interpersonal skills. The first axis corresponds to the level of technical or scientific knowledge as well as the mastery of technologies or processes. The second axis refers to the individual's interpersonal or managerial skills, especially team, project and relationship management skills. Each of these axes consists of several skills that are referenced in a competence sheet. Grading facilitates the distinction of roles that exist within a community by analyzing the level of excellence on each axis. In Figure 5.1, the letter "R" represents the generic roles that refer to all of the missions, knowledge and activities that may exist at these two-dimensional competence levels. The ellipses that surround these roles mean that the actual skills of those who perform such roles may be rather dispersed according to the specific aspects of generic roles. For example, in some professional communities, the roles of business manager and expert can be performed by the same person, which will not be the case in another community. Once the different roles are positioned, it is possible to link them with arrows, which indicate the possible paths within the specialty.

It has been shown that one of the bottlenecks in the management of monodisciplinary professional communities is dealing with role ambiguity and ensuring its evolution. "If we do not know the difference between a specialist and an expert, it is impossible to know when, how and in what conditions a specialist could become an expert" [LEL 14]. Mapping could thus become a useful tool for visualizing the structure of the professional community, allowing the enactment of mutually dependent roles and professional paths. The analysis of this map that reflects the composition of the community at a given moment should allow managers to then look forward to a "desired" state of the community and thus build a new "target". This could help technical and HR managers to coordinate their efforts to align knowledge management practices with the management of R&D professionals' roles and career paths.

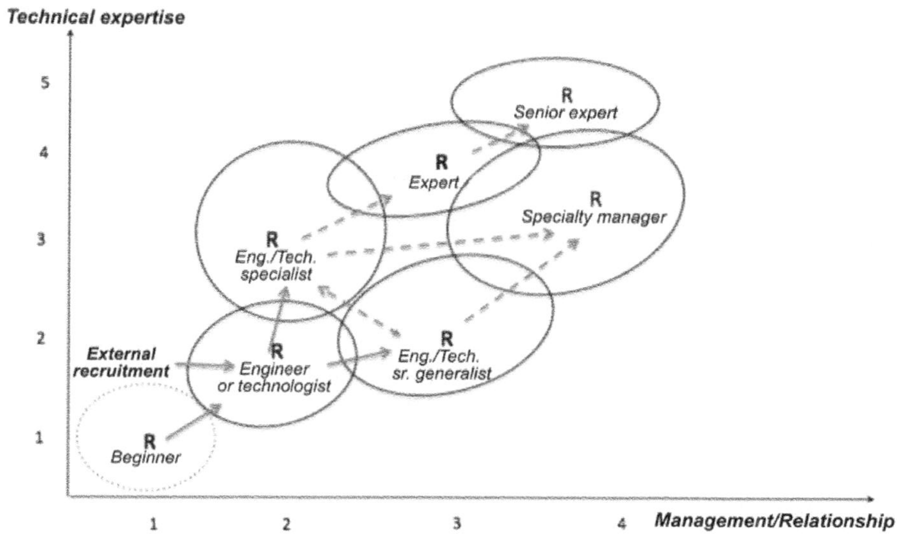

Figure 5.1. *Representation of a "monodisciplinary community target" in an engineering company. Source: [LEL 14]*

5.3. Conclusion

As we have seen, organizing collective action that could contribute to the strengthening of organizational expertise and innovation is imperative. This includes providing an environment that allows R&D professionals to invest in different communities of expertise and to be recognized and valued for their contributions. To create such environments, it is important for organizations to have the ability to distinguish and articulate the two forms of structuring of organizational expertise: horizontal and vertical.

Currently, it is observed that companies are increasingly demonstrating the need for transversality and adopting the "project" structure in particular. Their goal is to break with the rigidity of vertical structuring considered as a barrier for disruptive innovation. Within interdisciplinary communities, knowledge-sharing and the confrontation of ideas can lead to new interpretations and thus develop innovation. The role of transversal communities of expertise, such as R&D projects, troubleshooting teams and colleges of experts, seems to be crucial in this regard. However, cross-disciplinary exploration can only be effective if the company has strong specialized expertise. Indeed, too much transversality could disrupt learning and sharing dynamics, which is specific to the specialized monodisciplinary community. As explained above, the existence of such communities is strategic for developing high-level, company-specific expertise. As favored spaces of specialized knowledge and learning opportunities, these communities are crucial not only for organizational memory (they help to avoid "forgetting" critical expertise), but also for the development of new knowledge, which could be built on the solid foundation of professional expertise.

Because of their strategic nature, these communities thus become new management objects, requiring specific actions and tools, allowing them to align knowledge management challenges with competence and career management. Despite their importance, however, these communities have received little attention from decision-makers and managers. This lack of attention to such a strategic area of knowledge creation and maintenance could endanger, in the long term, the competitive competency of the company.

6

Performance Management in R&D

The concept of economic performance is important to a company. As Lorino [LOR 03, p. XI] recalls: "it involves creating value for customers, which includes meeting needs (by possibly generating new needs for innovation), under satisfactory conditions of cost, time and quality. Performance management incorporates all orientation and control activities aimed at achieving the desired level of 'performance'".

To address this subject, we first have to understand what the concept of performance and performance management is all about as well as identify performance challenges in research and development. This will be the purpose of the first part. To discuss what is meant by performance management in this area, we will then focus, more specifically, on R&D departments' management control, followed by that of innovation projects. We will conclude with reflections on the limits of rationalization of control in R&D.

6.1. Performance in R&D

6.1.1. *A hard to define concept*

In general, though performance is central in management, and perhaps because of that, this concept is yet to be given a single definition. Bourguignon [BOU 95] talks of the polysemy of the term "performance" and the debates around such a concept. Performance is success, but the representation of success is relative. What does success mean in R&D? In design activities, an invention does not always lead to an innovation (the positive sanction of the invention user), that is, for example, a more profitable process, a successful commercial transaction. And if the innovation is successful, who takes the credit for such a success?

Performance can be regarded as a combination of three criteria: efficiency (achievement of an objective), effectiveness (use of resources to achieve the objective) and relevance (comparison of objectives with allocated resources). The effectiveness of management by objectives is weak in the field of R&D. It certainly can focus on the use of allocated resources (compliance with budget), but we clearly see that this cannot represent the main aspect of this activity.

Performance can also be considered as a process, including putting skills into action. This view is more suited to R&D. Besides, it corresponds to the most recent orientations of management control and to the emphasis placed by some authors on process and competence-based management [LOR 03]. These two concepts are interdependent: “mastering action processes means having skills” [LOR 03, p. XI]. Inspired by this approach, it may simply be considered that if researchers, collectively, of a good standard and effectively cooperate with each other, as well as with downstream industry and marketing, then there is some potential for results to be profitable. This idea was put forth by the Japanese automobile industry in the late 1980s for its development activities, which is now referred to as “concurrent engineering”.

Beyond the generic difficulty in defining performance, there are some similarities to R&D which are attributable to a number of factors.

Firstly, we have the multiplicity of roles held, and there is often the tendency to reduce such roles to contribution to innovation projects. Meanwhile, R&D has several other missions, such as building a reputation of scientific credibility for the company, attracting human resources, judging irrelevant scientific and technical avenues, influencing standardization, guiding work carried out in the academic sphere and supporting the company to gain legitimacy with the public authorities.

It is difficult to maintain the concept of overall performance by adopting a balanced position between relevance, effectiveness and efficiency because said concept is not operational. This therefore leads some authors to distinguish between several types of performances. In the field of R&D and innovation, Hooge and Stasia [HOO 16, p. 50] distinguish four types:

– "frugal performance" aimed at optimizing R&D resources ("innovate more, spend less");

– "economic performance", equated with value creation;

– "strategic performance" that characterizes intangible and unquantifiable contributions; and

– "membership performance", in the sense of mobilizing the internal stakeholders of innovation.

We will discuss the first two types of performances when addressing the economic evaluation of projects (see section 6.3.1). Since the other two types of performances are qualitative in nature, they are more difficult to understand.

Regarding strategic performance, Hooge and Stasia [HOO 16] put forth a series of criteria. We will retain two among the most characteristic criteria: the level of consistency of R&D with the company's strategy and the contribution of R&D to communication and corporate image.

Membership performance is expressed in the manner in which R&D responds to innovation stakeholders' challenges and expectations within the company (project managers, technical teams, operational managers, purchase and marketing managers, etc.).

In addition to the issue of defining the different types of R&D performances, another difficulty lies in the often "invisible" nature of R&D results: in the technological field, for example, it is difficult to understand the influence or benefits of a major actor's reputation on the environment. How can we evaluate the benefits of research work which demonstrates that such technical avenue is a dead end and avoid misguided investments? Unfortunately, learning by failure [CUS 08] and learning from the failures of product or service innovation projects that did not meet with their expected success are practices that are neither widespread nor highly encouraged by the academic literature as it largely values success.

6.1.2. *Management difficulties specific to R&D*

There are many factors that make the management of R&D performance a complex exercise. Some relate generally to functional activities and others relate specifically to the field of design. This raises a series of issues.

– Issues of temporality. The issue of temporality is a tough one, in terms of both apprehending the results of R&D work and managing its activities. Indeed, we cannot evaluate the effectiveness of a management technique through the simple observation of immediate effects. The time lag between expenditures and the consolidation of benefits from the company's results can be very long. Thus, the dominance of short-term profitability over other valuation criteria severely penalizes breakthrough innovation projects that would generate a turnover in a more uncertain way and over a longer horizon. The pressure of short-term management can have perverse effects, as already discussed in the previous chapter.

– Issues related to the diversity of activities. Another factor that causes difficulties is the variety of activities covered by the term "R&D". At the very least, technical or tertiary supports, which already have a standard solution register, will be left aside. It is necessary to distinguish between "regulated design" activities covered by development and engineering and "unregulated design" (or innovative) activities that correspond to the research. In the first case, where it involves activating existing knowledge in accordance with predefined specifications, the quality/costs/period triptych remains the reference. In the second case, where it concerns producing new knowledge, the canonical model of performance is being questioned.

– A complex relationship between invested resources and recorded results. Research is characterized by the uncertain nature of its results. In particular, no modeling can establish convincing relationships between the resources mobilized and the economic results achieved (expanding an R&D department is not enough to make it more efficient).

– The issue of accountability. Today, managers agree with the idea that successful innovations of research make it more worthy than the inventions it manages to achieve. The problem is that the success (or failure) of an innovation can hardly be wholly attributed to R&D. This point is raised by Nixon [NIX 98, p. 332]: "one of the major difficulties in evaluating R&D activities lies in the fact that a successful innovation requires performance and structural characteristics that no single function can fully control". Much depends on the interactions between R&D and other corporate units and, as Nixon observes, interactivity is not constant in the innovation process. It evolves over time.

In general, R&D characteristics are inconsistent with traditional management control. Indeed, the latter is designed for operations in which activity, often organized over an annual timescale, is more or less standardized, with formalized processes and predefined objectives which are all controlled by a well-defined hierarchical structure [BER 16]. Also, despite the strategic importance of design and development activities, traditionally "control systems play a minor role at best" [GAU 02, p. 1]. However, performance challenges have not spared R&D, and this has led to renewed interest in the subject.

6.1.3. *Performance challenges*

The financial crisis has fostered attention focused on performance management in R&D which has for a long time been regarded as a discretionary activity, while management control specialists have attempted to provide (accounting and extra accounting) responses to the issue.

In addition, the shortening of product life cycles, increased competition and the accelerating pace of technological advancements (acceleration of product renewal, rapid product obsolescence due to corporate strategies and competitors, faster imitation) have all made the strategic context more tense. There is therefore a formidable scissor effect between the increase in product design cost on the one hand, and on the other hand, the drastic reduction of the period over which we could hope to achieve a return on investment.

The issue of efficiency also arises more clearly due to the relatively high wage structure of staff in Western countries and severe cost control pressures in the context of globalization. R&D is no longer spared by externalization. Since the early 2000s, large companies are no longer hesitant to set up research centers in China or India. In 2002, the cost of a Chinese R&D engineer was estimated at about 20% of that of a European or a US engineer. The gap has probably narrowed, but remains significant.

Other relevant specific sectoral issues add to such general performance issues. The following is an illustrative example in a changing industry: Pfizer, the world leader in the medicine industry (see Box 6.1).

Pfizer is a giant in the pharmaceutical industry, with a turnover of more than 53 billion dollars. In 2016, it was no. 1 in the industry and considered as one of the most innovative companies in the world (ranked 9th, in 2013, according to studies of the Booz & Company firm). The development of a new medicine is a long and costly process, takes roughly a dozen years, and is even made more complex by the health authorities' requirements and the concern to regulate the pharmaceutical expenditure of the States.

In 2006, Pfizer's R&D budget was 7.599 billion dollars for 10,500 employees in the industry. Like other companies in the sector, the company had to be subject to the expiration of patents of important molecules as well as generic competition, with a resulting considerable loss of revenue. Lipitor's (anti-cholesterol) patent expired in 2011, while it still represented 30% of Pfizer's sales in 2008. As the situation presented itself at the time, if no important acquired molecule was put on the market, Pfizer's revenues would have led to a decline in its turnover. Yet, estimated at 868 million dollars in 2006, the costs per molecule placed on the market were steadily increasing. Thus Pfizer had to respond accordingly.

In January 2007, the company published a plan to reduce 10% of the group's employees, and implemented a cost savings policy. Research sites were closed. The R&D model was undergoing immense changes. The group established public and private partnerships, including academic research, institutions and NGOs since it believed that the development of future medicines required external expertise. Other measures were put in place, for example, when faced with the loss of exclusivity with regard to Lipitor, the group launched its own generic to counter its competitors.

Box 6.1. *A challenged performance model: the case of Pfizer (source: Weinmann [WEI 08])*

6.1.4. *The delicate issue of measure*

In management, everything that is considered important is supposed to be subject to a measure. This also applies to performance. This concern is expressed through key performance indicators, mostly referred to by the acronym KPI. These indicators (rates, quotients, percentages, averages) are grouped in a scorecard intended to be used in the performance management of the field taken into account and including appropriate explanations and comments for the conduct of future actions.

6.1.4.1. *Classical indicators*

In France in 2015, a survey by questionnaire and interviews conducted by the Université de Technologie de Compiègne and the KL Management firm [FUG 16] with a sample of 90 managers of R&D-related functions of industrial companies from different industries (aeronautics, automotive, chemistry, pharmaceutical, etc.) shows that barely 10% of these companies follow indicators specific to the R&D function. These companies focus mainly on projects and mostly use the following five indicators:

– observation of the project's internal milestones;

– respect of launching dates;

– profitability of the project;

– control of the project budget; and

– keeping with individual objectives.

Taking into consideration international practices, according to Gallié *et al.* [GAL 10], who summarized the international literature, three main indicators are used to measure the tangible results of R&D: patents, publications and innovation.

Patents are the most common knowledge production measure. This measure has several limitations. Not all inventions are patentable (some intellectual creations are excluded). All knowledge is not codifiable; this is what we call tacit knowledge. Although such knowledge cannot be stored in databases, they are, however, often the most crucial. Finally, patents are costly (regarding their filing and maintenance) and may turn out to be quite ineffective in protecting an invention, and moreover, some companies may prefer secrecy.

Research activity can also be measured by the number and prominence of academic publications. This is the basic indicator in assessing the activity of public research structures. In the private sector, mainly for high technology activities and upstream research, researchers can also be assessed on the basis of this criterion. But this type of measure relies on peer-controlled recognition mechanisms and delivers knowledge to the public domain. Yet, it is not well known to the company, which is therefore not very willing to use it.

The third type of indicators is innovations. They express the R&D result of inventions placed on the market that trigger purchase actions from a particular consumer segment. They specifically constitute economic characteristics and as such are naturally valued by organizations. However, there are different views to the definition of an innovation, and this concept is far from corresponding to a homogeneous reality. In addition, accepting innovation as such ignores process innovations that are developed internally and thus contribute indirectly to economic performance.

As noted by Fugier-Garrel *et al.* [FUG 16, p. 35], curiously, customer satisfaction indicators are not in the top five most used KPIs. More broadly, some researchers have suggested that performance assessment should be extended to a wider range of internal and external actors by further entrenching control within the company's strategy. This is the purpose of the balanced scorecard (BSC), proposed by Kaplan and Norton [KAP 96]. In R&D, the BSC aims at providing a relevant response to the expectations of each of our innovation stakeholders.

6.1.4.2. *Taking the various stakeholders into account*

Bremser and Barsky [BRE 04] use the BSC as an integrating framework to show how companies can link their resource commitments in R&D to the company's activities and objectives. The purpose of this approach is to contribute toward a logic of matching the objectives of groups of actors with strategic objectives. Researchers provide some guidelines on how companies can apply this integrated performance design to the R&D measure system.

In the illustration provided (based on the analysis of a research center's strategy) the authors propose four "strategic objectives" which are the perspectives of different stakeholders: financial, customer, internal professions and employee growth and development perspectives. There are KPIs corresponding to these objectives throughout the entire company. For example, for the development perspective, we have employee retention, their development, their organizational climate survey, etc. Metrics are in turn linked to one or more KPIs. For example, Bremser and Barsky suggest a skill coverage ratio by a strategic skill category, linked to the "employee retention" indicator, or the calculation of the volume of training hours, linked to the "retention" and "development" indicators.

Taking multiple stakeholders into consideration can lead to some complexity, as shown in the study by Agostino *et al.* [AGO 12] on an Italian technology research center. As part of an action research aimed at developing an R&D performance measure system intended to balance the diversity of expectations, they propose a set of 23 indicators grouped into five dimensions. An overview of this measure system is provided in Table 6.1, using 12 of the key indicators cited.

DIMENSION	KPI	STAKEHOLDERS			
		Admin. manager	**Scientific director**	**Executive committee**	**Research unit director**
Efficiency	Number of scientific publications	X	X	X	
	Number of patents	X	X	X	
	Stocks of competitive projects		X	X	X
Result	Revenue generated by patents, licenses and other products	X	X	X	
	Revenue generated by technology transfers	X	X	X	
Effectiveness	Cost of a scientific publication	X		X	
	Patent management and maintenance cost	X	X		
	External/internal budget	X	X	X	
Risk	Scientific reputation	X	X	X	X
	Turnover		X	X	
	Organizational climate				
Network	Number of joint publications		X		X

Table 6.1. *Example of KPI according to stakeholders' priorities (excerpt from Agostino* et al. *[AGO 12, p. 56])*

It is very difficult to discuss the relevance of the proposed indicators since they relate to a specific organizational context and research process. It is the concern for interaction with stakeholders that has peaked our interest, just as in the case of the intrinsic value of KPIs.

Following the distinction proposed by Simons [SIM 90], between diagnostic control (the classic conception of management control) and interactive control (based on horizontal interactions between members of the project team and vertical interactions between managers and subordinates), Gautier [GAU 02] advocates the importance of an interactive control, speculating that this approach is more suited to development projects insofar as it makes it possible to address the uncertainty inherent to projects. The use of the BSC could be considered in some respect as an extension of this reasoning, to the extent that it can be designed and used in a participatory and transversal approach to foster interactions between different groups of actors involved in the different phases of projects.

6.1.4.3. *The limits of technical sophistication*

Innovation performance has been extensively studied. However, so far, the results of the many studies have not produced a unanimously recognized performance indicator. The diversity of indicators according to industries, countries and the type of measures is considerable. Since the end of the 1990s, there has been a proliferation of indicators and a certain sophistication of control systems, but the results have not been very convincing.

The study by Hagedoorn and Cloodt [HAG 03] on the technological performance of 1,200 companies in several industries (aeronautics, pharmaceuticals, etc.) shows that, on the one hand, these indicators tend to reinforce each other and, on the other hand, each of them taken alone provides a relatively satisfactory approximation of innovation performance. The results of this study point toward a certain pragmatism.

The BSC, however, may represent a way forward, provided, it seems to us, that we hold on to the proposal of Naro and Travaillé [NAR 11, p. 78] who posit that far from a normative and universal approach, "each organization must find its own way, even if this means shaping and twisting Kaplan and Norton's original model according to its identity, strategic vision and especially representations and interactions between the actors involved".

6.2. Budgetary control of R&D departments

All companies running an R&D activity do not necessarily have a department dedicated to such an activity. However, in large companies, this is most often the case, except for those that position R&D as their core activity, as is the case, for example, in advanced electronic activities which are organized in a matrix manner, crossing two divisions: by geographical area and by activity sector.

As elsewhere, budget management is a resource allocation mechanism related to the funding of an activity, and as such, it is an essential part of management control. The budget, which represents the cost of a program, as part of a plan, aims to clarify all of the forecasts that are considered as standards to be respected. Let us look at each of these key terms: plan, program and budget:

– the plan focuses on providing what should be the R&D department's activity over a multi-year period. It is expressed in qualitative and quite general terms by setting major guidelines for the future;

– unlike the plan, the program is the short-term forecast (often for a year) specifying in detail and taking different contingencies, activity levels, volumes agreed (workforce particularly) into account, etc.; and

– the budget represents a translation of the plan, by calculating the implementation of the program defined on different cost factors in monetary units.

The R&D department's costs mainly cover salaries, materials used and equipment needed for this activity [BER 16]. Their respective share greatly varies according to activity sector, for example between digital and more capital-demanding areas that mobilize costly equipment such as the steel industry.

Wages and related costs often represent the most significant item. Determining the exact number of staff required (in number and qualification) is the necessary prerequisite for this calculation. Depending on the activity, equipment for trial, testing and experimentation may equally represent an important item. Premises also represent a clearly identifiable item. Finally, it is customary to include the expenses necessary for technological observations in an R&D department's expenditure. This could include subscriptions to journals or participation in various scientific or professional events (congresses, conferences, etc.).

Alongside such expenses, the R&D department can also generate certain products: revenues from royalties or patent dedications, securing grants (notably State or European Union funding). They can also provide paid services to external actors (contract research, sample characterization, etc.).

By integrating all these factors, the outlines of an R&D department's budget can be presented as listed in Table 6.2.

	Items	**Amount**
EXPENSES		
Staff costs	Net wages, bonuses and social charges, remuneration-related payments (vocational training, taxes)	
Remuneration of non-permanent staff	Purchase of research services, temporary staff	
Equipment	Amortization of computer equipment and others, equipment rental cost	
Premises	Rental fees	
Technological observations	Print or online journal subscriptions Costs of participation in congresses, conferences and seminars	
Miscellaneous costs	Travel expenses, energy consumption, cost of supplies and other consumables	
TOTAL EXPENDITURE		
VARIOUS RESOURCES		
Subsidies	Public funding of equipment, operation or investment	
Patent products	Assignment price or royalties	
External services	Contract research, paid expertise, provision of equipment, etc.	
TOTAL REVENUE		

Table 6.2. *Example of an R&D department budget*

The budget, apart from its contribution to management control, in the narrow sense, also plays two other roles: in its design phase, the budget plays the role of a simulation and decision support tool, and in the course of the budget cycle, that of motivation and conflict management. The last point is central in an activity where technological innovation imperatives can interfere with the objectives of financial equilibrium.

6.3. Innovation project management control

Project management can be considered at two levels: that of the entire company that generally manages a project portfolio and that of the project manager who manages the project "internally". Project portfolio management focuses specifically on projects for the purpose of selection and mediation. This is the subject discussed in section 6.3.1. With regard to the project manager, this involves a proper conduct of the project. This issue will be addressed in section 6.3.2.

6.3.1. *Economic assessment of projects: the two approaches*

Despite the generalization of project management, project selection and assessment remains a major concern for companies, due to a high failure rate and uncertainty that limits the use of traditional assessment and control methods. The design activity field is obviously not left out of this customary practice. Here, cost and time abuses are usual, especially because the activities of this field are less predictable and programmable than other activities of the company.

To remedy such a practice, approaches have been developed that can be divided, in accordance with Garel [GAR 03a, GAR 03c], into two broad categories: cost and profitability approaches.

6.3.1.1. *Cost assessment*

Much of the literature on project management especially covers this subject and distinguishes several cost concepts. According to Garel [GAR 03a, GAR 03c], whose terminology we have adopted, there are three main types of project cost assessment to which three forms of design correspond: "design to value", "design to cost" and "design to target".

In "design to value" methods, the project is launched on the basis of the appreciation of a client's need. A specification is drawn up from which technical specifications are derived and, based on cost accounting, a cost calculation is carried out.

In "design to cost" methods, which are part of the competitiveness optimizing approaches, determining cost is no longer based on the specifications but on the profit requirements of the company, in accordance with the customer's expectations (quality, reliability, willingness to pay, etc.). Here, cost constraint is crucial and extends from the product's design phase (target costing) to its entire life cycle. This approach, which also aims to reduce design periods, is now greatly used by companies.

An emblematic illustration of this second approach is that of Renault's Logan, a "real" profitable car at € 5,000. The design to cost method that was introduced in the company in 1992, with the Twingo project, grew considerably with the launch of the Logan project in 2000. The design process was focused primarily on cost savings through the use of proven solutions (see Box 6.2).

"The idea was to resolutely guide the entire design process in a reliable and economic approach by using proven solutions. The methodology integrated elements of product and process analysis. The parameters included feasibility, expected gains, local availabilities and, the primary criterion, reliability. Thus, with regard to materials, they decided to use conventional steels, which properly adapts to the methods and means of production of the planned manufacturing sites, with their less robotized processes, which could be "sourced" locally. Similarly, from the outset, Logan's design integrated stamping constraints by limiting the presence of body edges to facilitate the creation of manufacturing tools, increase the reliability of the stamping and sheet metal work process, and reduce the cost of these operations. The low convexity of the glazing, especially the rear window, reduces the cost of tools [...] Other significant gains were obtained through the forming of the brake lines at the factory, their protection by stringers without any additional spare parts inserted or a tank with integrated pipe and removable external filter for countries using "unstabilized" fuels. Another key element behind the success of Logan's economic gamble is the choice of a front end similar to that of a Clio without an anti-roll bar and a rear end taken from the Renault-Nissan Alliance B platform. Moreover, to reduce investment in tooling and by simplify the assembly, the mirrors and bumpers are designed to be placed interchangeably on the left or right side. Finally, a single window on the rear doors saves the cost of an inserted window [...] The use of the same elements in several vehicles of a car manufacturer provides assurance regarding reliability for customers and guarantees economy for the brand."

Box 6.2. *Logan: a breakthrough innovation, based on design to cost methods (source: excerpt of Renault press release, published on June 2, 2004)*

Authors have criticized some overly static approaches to design to cost methods. In particular, Gautier and Giard [GAU 00] argued that calculating a cost at a given moment without taking into account changes over the product life cycle was a weakness of the method. "Design to target" methods address this criticism.

Design to target methods consist of the flexibility of the "objective cost". Garel [GAR 03a, GAR 03c] gives the example of household appliances, microcomputing or mobile telephony, where the objective cost of a product can be adjusted several times in the course of its development, depending on risks (introduction of a competing product, etc.) or business opportunities (opening of a niche for export, etc.).

6.3.1.2. *The assessment of financial profitability*

Financial profitability approaches were first developed in response to the question "should a project be launched or not?" but these approaches can still be mobilized well after the launch of the project as long as improvements and adjustments can be made to the project's cost. Some research carried out in Japan has also highlighted the importance of delaying certain decisions as long as possible to improve profitability.

A first approach, the most conventional, considers projects as investments that must yield fruits, and it is based on cash-flow capitalization, by the net present value (NPV). A project is accepted and funded (a) if the NPV of positive cash flows exceeds the NPV of the negative cash flows and (b) if the NPV is higher than that of another project with which it is competing for the allocation of resources. However, though an R&D project can be associated with a sequence of expenditures followed by revenue, in innovative design, there is high uncertainty regarding the figures on which expenditure analysis is based, let alone for future revenue, concerning both their amounts and their implementation schedule; thus, a calculation rationality seems quite illusory.

Ultimately, "an investment project can be launched despite a negative NPV, because the top management believes that it is strategic in nature. Conversely, a project with an interesting NPV may be rejected if it does not fall within the general strategy of the company" [GAR 03c, p. 80]. The second case does not always produce an eclipsing of ideas that can lead to a spin-off, which is how the creation of a new company is referred to, a start-up that will ensure the economic valuation of know-how and research results. Such a decision, by making an activity considered non-strategic independent, will provide project owners with a framework that is more conducive to the development of their ideas and a greater readability of accounts to investors. We can cite, for example, Arculus, which was created by Audi car manufacturer, to allow some of its engineers to work on concrete applications of flexible and reconfigurable factory concepts, developed in-house but inapplicable as it is in its historic factories.

A second, more recent approach is based on the options theory that was developed in the early 1970s for financial markets. It should be recalled that an option is a right to buy (call option) or sell (put option) an asset (bonds, shares, currencies, etc.), at a pre-specified price, at any time of a predetermined period. This theory, which aims at optimizing strategic decision-making by analyzing, over time, its related risks and opportunities, has been mobilized to value non-financial assets such as capital goods, production units, or R&D projects. This is referred to as a “real option” (option of abandonment, investment backlog, modification of its intensity, etc.). Risk and opportunity assessment is conducted through estimated profitability calculations.

Real option analysis has often been recommended as a suitable method for assessing high-risk investment projects, as may be the case for technological innovation projects with very high failure rates: according to an article published in CNRS journal, on May 21, 2015, seven to nine innovations out of ten witness failure. Some abandoned innovations can also be reborn. Hence, the project “Archeology of abandoned, neglected or emerging innovations” (ARIAD) was launched by the CNRS to understand the reasons for abandonment, failure or recovery.

The real option method was mobilized particularly in the biotechnology [MIL 15] and pharmaceutical [HAR 06a] fields. In these two sectors, the success of a company depends largely on R&D, and the main challenge is allocating resources to the best science projects.

In his doctoral dissertation, Krychowski [KRY 07] also presents case studies on actual options applied to decision-making in R&D in the telecommunications sector. Krychowski emphasizes in particular that methodology is an effective tool for a better understanding of the value of an investment project in a context of uncertainty. She relies on the case of the deployment of an ADSL (asymmetric digital subscriber line) network in a sparsely populated area (which includes uncertainty about the profitability of the project, despite public subsidies).

Thus, NPV or real options? Real option models are complex to apply, difficult to explain to an industrialist and therefore lack transparency. Ultimately, we note that the financial and strategic analysis areas are not always reconcilable. Moreover, the real option concept, after having generated a strong theoretical interest, is less appreciated today. Jacquet and Philippe [JAC 06] in this respect talk of a game of "smoke and mirrors". Other techniques have been developed to use the NPV in situations of uncertainty: decision trees in particular, but also simulation when the uncertain can be partially probabilized.

6.3.1.3. *Cost and profitability are not the only decision criteria*

Though the evaluations of the assumed costs and benefits of a project are rational criteria that come spontaneously to mind, the company, on the basis of many other criteria, accepts or rejects a project, as well as establishes an order of priority among several projects, based on the resources that it is likely to allocate. Berland and de Rongé [BER 16] talk of:

– consistency with the company's strategy;

– synergy with existing technologies;

– competences available within the company;

– the company's ability to patent the object of innovation; and

– project compatibility with existing production systems.

These reflections fully justify, as we have already mentioned, why R&D performance should not be reduced to its quantitative dimensions alone. No algorithm can guarantee the right choice of project.

6.3.2. *Project management methods and tools*

6.3.2.1. *Managing projects*

Project form, which was originally limited to engineering (space industries, construction, defense, nuclear, etc.), generated considerable interest, and since the 1980s, it has spread in all sectors of activity. In R&D, project-based organization is widely used to promote the innovation process.

Once the decision to launch an innovation project is made, such a project has to be managed. This then comprises the object of project management which includes various aspects. Methodologies focus mostly on time and budget. We believe it is necessary to consider human resources, which are also a key element for project success. With regard to HR methods concerning R&D staff, we refer the reader to Chapter 4.

6.3.2.2. *Project structuring and time management*

Time management tools (planning and project progress monitoring) supported by specialized software packages that help to disseminate them and used to check the proper performance of various planned tasks are the most common: Work Breakdown Structure (WBS), the Gantt Chart, the PERT Chart, etc.

The WBS provides a visual representation of the work to be done, breaking down the project into batches of successive works to a level of detail aimed at a meticulous assignment of tasks, careful planning and effective monitoring of the operations to be performed. It is some sort of workflow that serves as a basis for each level of the WBS that presents workflow up to the final level and that of the concrete tasks to be carried out. Each batch is assigned to an entity and a different manager.

Depending on the objectives, other types of scheduling structures are possible. The Resource Breakdown Structure (RBS) splits the project into resources and groups these resources by nature or work teams. The project manager, through the RBS, sets up work schedules, assigns duties and checks availability in terms of skills. The project manager can also structure the project according to profession or qualification. The RBS makes it possible to identify the different profiles required for the project and also to anticipate the necessary needs (recruitment, mobility and training). Through a crossing of an RBS (horizontally) and a WBS (vertically) in a matrix, it is possible to visualize the professional groups that would be in demand relative to the tasks to be performed and to check that all of the project's tasks have a sufficient number of resources and sufficient qualification to facilitate their execution (see Table 6.3). Such analysis may be particularly relevant in a multi-project context.

PROJECT X	**RBS (in days)**				
	From April 9–30, 2018	Engineers	Technicians	Workers	Total
	Batch 1 (April 9–14)	*25*	*19*	*5*	*49*
	Including task A	17	4	/	21
	Task B	8	15	5	28
WBS	Batch 2 (April 16–30)	*32*	*19*	*7*	*58*
	Including task C	12	6	2	20
	Task D	10	5	2	17
	Task E	10	8	3	21
	Total (batches 1+2)	*57*	*38*	*12*	*107*

Table 6.3. *Assignment of resources to tasks*

The Gantt chart was designed in the early 20th Century by a mechanical engineer, Henry Gantt. It consists of a list of ordered tasks (on a vertical axis) facing a schedule (on a horizontal axis). To construct this diagram, we use the WBS tasks list that we superimposed to the schedule. Thus, finally on the same diagram, we have the different tasks to consider, the start and end date of each task, the expected duration of each task, the possible duplication of tasks and the period of such duplication as well as the project start and end dates. The Gantt chart is easy to understand and can be quickly obtained through planning software. In counterpart to such simplicity, the analysis it provides is poor, in particular it does not highlight the critical tasks of a project.

The PERT method (Program Evaluation and Review Technique) was developed for the US Navy in the 1950s. It can be considered as a sophisticated variant of the Gantt chart. Unlike the latter, with the PERT, the critical path of a project can be determined, which is the longest sequence of tasks that must be performed for the project to be completed on the due date. The PERT method is more complex than the Gantt chart and as such is synonymous with complex and long-term projects.

6.3.2.3. *Project management control*

In the extension of a project structure by tasks and resources, the Cost Breakdown Structure (CBS) organizes a breakdown of the project by a budget envelope. The budget is calculated by consolidating tasks at the level of the entire project and increasing such to cover unforeseen expenses. The preliminary budget made to estimate the distribution of the planned expenditure of the project is followed by a reference budget for the follow-up of the project. Classically, starting from a table like the one previously presented, the other direct (specific equipment, purchases, etc.) and indirect (administrative follow-up, financial expenses, etc.) costs are multiplied by a daily rate established on the basis of the average gross wages allocated to each batch. Various methods can be used to estimate workloads and related costs, including load assessment with reference to an older project (analog method).

Cost control, known as "cost containment", includes all activities that aim at maintaining expenditure within the budget envelope (following the Afnor FD X 50-137 standard). Garel [GAR 03c, p. 80] distinguishes three types of costs:

– the budgeted cost of work scheduled (BCWS) and the expected cost of the project corresponding to the expected progress. This value results from the negotiation between the project manager and the business actors for the work to be done;

– the actual cost of work performed (ACWP) at a given date, which establishes what really happened. It is the monetary expression of the amount of resources committed to carrying out the work. It requires the translation of the use of equipment and the number of hours worked into monetary units, which is in line with the cost accounting conventions of the company;

– and the budgeted cost of the work performed (BCWP) on that date, also called the acquired value. This is the budget value of the completed project, based on the agreement between the project actor and the business actors.

Other notions like invoiced costs can also be taken into account. Various performance indicators ensue from analyzing the differences between these costs. For example, the cost variance measures the difference between the budget and what is realized at a given date. Thus – provided that the unit costs of resources remain unchanged – if the gap between BCWP and

ACWP is zero, then the work has been completed with the productivity that was initially forecast. If it is positive, then the work was carried out with a higher than expected productivity. With the same reservation, if the difference is negative, then the work was carried out with lower than expected productivity. The cost difference can be explained by changes in the work required, for example, unexpected difficulties can lead to delays and an increase in work hours, or changes in the cost of resources, for example, soliciting a new and more expensive supplier, considering the poor quality of the habitual supplier, or the need for incidental tests resulting in the additional use of expensive equipment.

The progress of the project (cost and schedule tracking) is graphically represented to provide visual information on the economic situation and the calendar of activities for the project. This is the so-called "S curve" (see Figure 6.1). The project follows this form with a gradual kick-off, followed by a speed-up, then by a slow down as the project gets to completion. The abscissa show the time and the ordinate shows the cumulative costs. On the abscissa line, the date on which the analysis is carried out (J), the contractual project end date (Jc) and the re-estimated final date (Jp) is indicated. The ordinate line shows the budget on date J (BD) and the re-estimated provisional cost (PC).

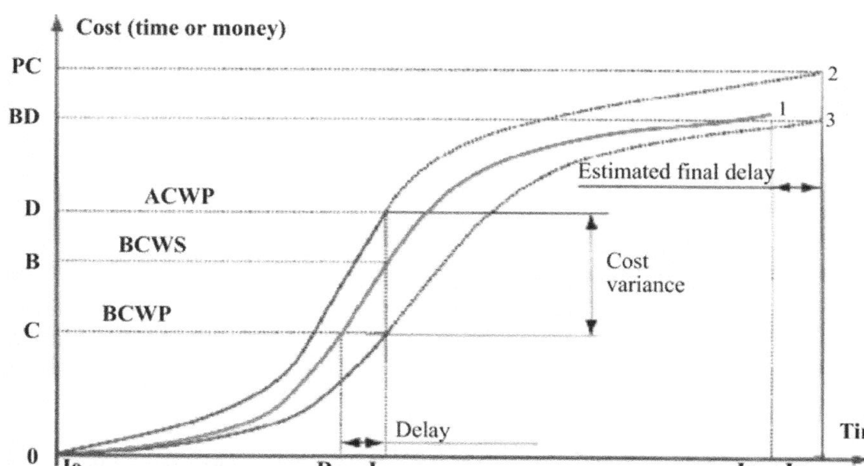

Figure 6.1. *Allocation of resources to tasks*

The first curve drawn (in 1) is the "budgeted cost of work scheduled" curve (BCWS). The second curve (in 2) is that of the actual cost of work performed (ACWP). The third curve (in 3) is that of the budgeted cost of work performed (BCWP).

The method makes it possible to compare, on the spot, what was planned to be done and spent with what was actually achieved and at what cost. In the event of budget overrun or delay, corrective actions will be taken to limit excesses.

6.3.2.4. *The advantages (and limitations) of rationalization*

Just like other North American specialists, Larson and Gray [LAR 10], coauthors of a bestseller on project management, lay great emphasis on the pros of IT in measuring project performance, asserting that it is indispensable for effective control. It is true that the computerized means of data collection have significantly improved the formatting and the analysis of data for the duration of a project and thus facilitated its monitoring and detection of problems. However, the contribution of management tools to control should not be exaggerated.

François Jolivet [JOL 03, p. 81], a practitioner who has managed many construction projects, assumes that the manufacturing company is still characterized by a "mechanistic vision": "the implicit reference of managers and production: they would like to codify, rationalize, standardize, and mechanize the process of design and the development of new products, in the same way as production [...] One needs to have handled or analyzed many projects to know that these management tools are not associated with performance".

Rationalization or agility? Today, this appears to be the main dilemma of the project management that Triomphe [TRI 13] formulates in her text on this issue. Like Triomphe, we believe that it is not about taking a radical position with respect to these two terms but to rather bring them together to remain efficient and get new contracts.

In the traditional conception of project management, the doctrine of industrial rationalization is still very strong. However, while it can be adapted to incremental innovation projects, it is counter-productive in the design that aims for breakthrough innovation and that is not so much concerned with improving the design of exiting products, but rather with exploring new avenues for product creation. It will therefore be risky to follow the tendency to formalize the design activity.

To resolve this dilemma, researchers at the Ecole des Mines de Paris [HOO 16, p. 126] name specific tools for managing value creation, emphasizing that tools not only have a role of rationalizing but also of reconditioning innovation: "they downplay the debate by shifting it from a discussion on the subjective and emotional perception of the potential of innovation to a debate on collective ways of actions for optimizing innovative design activity" .

6.4. Conclusion

Like other authors who preceded us in studying this thorny subject, we cannot but note the dilemma facing the company. On the one hand, we recognize that it would be risky to rely exclusively on common sense and the spontaneous adjustment of actors' willingness to manage R&D activities. On the other hand, it seems no less obvious to us that by wanting a more sophisticated traditional management control in order to adapt to the complexity of the field, we may end up stifling and impeding innovation in an attempt to control it. The issue therefore involves aiming for a harmonization between autonomy and control.

These observations point to a broader and humble approach to performance management, in line with Chiapello [CHI 96, p. 52], according to whom control is "any influence that creates order, that is, a certain regularity".

Bibliography

[AGO 12] Agostino D., Arena M., Azzone G. *et al.*, "Developing a performance measurement system for public research centres", *Journal of Business Science and Applied Management*, vol. 7, no. 1, pp. 43–60, 2012.

[AKH 14] Akhilesh K.B., *R&D Management*, Springer, New York, 2014.

[ALL 77] Allen T.J., *Managing the Flow of Technology*, MIT Press, Cambridge, 1977.

[ALL 86] Allen T.J., Katz R., "The dual ladder: motivational solution or managerial delusion?", *R&D Management*, vol. 16, no. 2, pp. 185–197, 1986.

[ALL 08] Allouche J., Domenach J.-L., Froissart C. *et al.*, Les entreprises françaises en Chine : environnement politique, enjeux socio-économiques et pratiques managériales, Centre d'études et de recherches internationales (CERI), Institut d'Etudes Politiques de Paris, 2008.

[ALV 05] Alvarez J., Svejenova S., *Sharing Executive Power: Roles and Relationships at the Top*, Cambridge University Press, Cambridge, 2005.

[AMA 83] Amabile T.M., "The social psychology of creativity: a componential conceptualization", *Journal of Personality and Social Psychology*, vol. 45, pp. 357–376, 1983.

[AMA 89] Amabile T.M., Gryskiewicz N.D., "The creative environment scales: work environment inventory", *Creativity Research Journal*, vol. 2, pp. 231–252, 1989.

[AMA 96] Amabile T.M., *Creativity in Context*, Westview Press, Boulder, 1996.

[AMA 99] Amabile T.M., Conti R., "Changes in the work environment for creativity during downsizing", *Academy of Management Journal*, vol. 42, no. 6, pp. 630–640, 1999.

[AND 72] ANDREWS F.M., FARRIS G.F., "Time pressure and performance of scientists and engineers: a five-year panel study", *Organizational Behavior and Human Performance*, vol. 8, pp. 185–200, 1972.

[ART 89] ARTHUR M.B., HALL D.T., LAWRENCE B.S. (eds), *Handbook of Career Theory*, Cambridge University Press, Cambridge, 1989.

[ASQ 07] ASQUIN A., GAREL G., PICQ T., "Quand les individus et les collectifs sociaux sont mis en danger par le travail en projet", *Gérer et Comprendre, Les Annales des Mines*, pp. 43–54, 2007.

[BAD 13] BADILLO P.Y., "Les théories de l'innovation revisitées : une lecture communicationnelle et interdisciplinaire de l'innovation ? Du modèle 'Emetteur' au modèle communicationnel", *Les enjeux de l'information et de la communication*, no. 14/1, pp. 19–34, 2013.

[BAK 05] BAKER T., NELSON R.E., "Creating something from nothing: resource construction through entrepreneurial bricolage", *Administrative Science Quarterly*, vol. 50, pp. 329–366, 2005.

[BAR 86] BARON R.S., "Distraction/conflict theory: progress and problems", in BERKOWITZ L. (ed.), *Advances in Experimental Social Psychology*, Academic Press, New York, 1986.

[BAR 91] BARLEY S.R., TOLBERT P.S., "Introduction: at the intersection of organizations and occupations", *Research in the Sociology of Organizations*, vol. 8, pp. 1–13, 1991.

[BAR 99] BARON X., "Gestion des ressources humaines et gestion par projet", in WEISS D. (ed.), *La Fonction ressources humaines*, Éditions d'Organisation, 1999.

[BAS 13] BASADUR M., GELADE G., BASADUR T., "Creative problem-solving process styles, cognitive work demands and organizational adaptability", *Journal of Applied Behavioral Science*, vol. 50, no. 1, pp. 80–115, 2013.

[BEA 12] BEAUJOLIN R., SCHMIDT G., *Restructurations d'entreprises*, La Découverte, Paris, 2012.

[BEN 98] BEN MAHMOUD-JOUINI S., Stratégies d'offres innovantes et dynamiques des processus de conception : le cas des grandes entreprises générales de bâtiment françaises, PhD thesis, Paris Dauphine University, Paris, 1998.

[BEN 06] BEN MAHMOUD-JOUINI S., "Conjuguer management des projets et management des connaissances", in BOUNFOUR A. (ed.), *Capital immatériel : connaissance et performance*, L'Harmattan, 2006.

[BEN 16] BEN MAHMOUD-JOUINI S., "Le numérique au service des entités dédiées à l'innovation de rupture", *Revue française de gestion*, vol. 1, no. 254, pp. 65–87, 2016.

[BÉR 03] BÉRET P., MENDEZ A., PARAPONARIS C. *et al.*, "R&D personnel and human resource management in multinational companies: between homogenization and differentiation", *International Journal of Human Resource Management*, vol. 14, no. 3, pp. 449–468, 2003.

[BER 16] BERLAND N., DE RONGÉ Y., "Le contrôle de la R&D et de l'innovation", *Contrôle de gestion : Perspectives stratégiques et managériales*, Pearson, Paris, 2016.

[BLU 92] BLUERDORN A., KALLIATH T., STRUBE M. *et al.*, "Polychronicity and the inventory of polychronic values (IPV): the development of an instrument to measure a fundamental dimension of organizational culture", *Journal of Managerial Psychology*, vol. 14, no. 3/4, pp. 205–231, 1992.

[BOB 14] BOBADILLA N., A longitudinal study of the effects of slack reduction on creativity in R&D and creative teams, PhD thesis, Pantheon-Sorbonne University, Paris, 2014.

[BOB 15] BOBADILLA N., GILBERT P., "Managing scientific and technical experts in R&D: beyond tensions, conflicting logics and orders of worth", *R&D Management*, vol. 46, no. S1, pp. 1–13, 2015.

[BOM 99] BOMMER M., JALAJAS D., "The threat of downsizing on the innovative propensity of R&D professionals", *R&D Management*, vol. 29, no. 1, pp. 27–34, 1999.

[BOU 81] BOURGEOIS L., "On the measurement of organizational slack", *The Academy of Management Review*, vol. 6, no. 1, pp. 29–39, 1981.

[BOU 95] BOURGUIGNON A., "Peut-on définir la performance ?", *Revue française de comptabilité*, no. 269, pp. 61–66, 1995.

[BRA 93] BRABET J., *Repenser la gestion des ressources humaines ?*, Économica, 1993.

[BRE 04] BREMSER W.G., BARSKY N.P., "Utilizing the balanced scorecard for R&D performance", *R&D Management*, vol. 34, no. 3, pp. 229–238, 2004.

[BUR 61] BURNS T., STALKER G.M., *The Management of Innovation*, Tavistock, London, 1961.

[CAB 17] CABANES B., Modeling the emergence of expertise and its governance in innovative organizations: from communities to proto-epistemic societies of experts, Thesis, Mines ParisTech – PSL Research University, 2017.

[CAL 11] CALAMEL L., DEFELIX C., MAZZILLI I. *et al.*, "Les pôles de compétitivité : un point de rupture pour la GRH traditionnelle ? Une analyse des dispositifs RH au sein des 12 pôles de la région Rhône-Alpes", *Management & Avenir*, vol. 41, no. 1, pp. 175–193, 2011.

[CAM 94] CAMERON K.S., "Strategies for successful organizational downsizing", *Human Resource Management*, vol. 33, no. 2, pp. 189–211, 1994.

[CHA 97] CHARUE-DUBOC F., "Maîtrise d'œuvre, maîtrise d'ouvrage et direction de projet, pour comprendre l'évolution des projets chez Rhône-Poulenc", *Gérer et Comprendre, Les Annales des Mines*, pp. 54–64, 1997.

[CHA 01] CHARUE-DUBOC F., MIDLER C., "Développer les projets et les compétences – Le défi des hiérarchiques dans les métiers de conception", *Gérer & Comprendre, Les Annales des Mines*, no. 63, pp. 12–22, 2001.

[CHA 05] CHANAL V., DEFELIX C., GALEY B. *et al.*, "Les personnes innovantes dans les entreprises doivent-elles faire l'objet d'une GRH spécifique ? Une étude exploratoire", *Gestion 2000*, vol. 2, pp. 99–113, 2005.

[CHA 12] CHARUE-DUBOC F., GASTALDI L., "Recherche et développement. Le management des professionnels de la R&D", in ALLOUCHE J. (ed.), *Encyclopédie des ressources humaines. Théories, instruments, méthodes, auteurs*, 3rd ed., Vuibert, Paris, 2012.

[CHA 15] CHANDRASEKARAN A., LINDERMAN K., SCHROEDER R., "The role of project and organizational context in managing high-tech R&D projects", *Production and Operations Management*, vol. 24, no. 4, pp. 560–586, 2015.

[CHE 03] CHESBROUGH H.W., *Open Innovation: The New Imperative for Creating and Profiting from Technology*, Harvard Business School Press, Boston, 2003.

[CHE 06] CHESBROUGH H.W., "Open innovation: a new paradigm for understanding industrial innovation", in CHESBROUGH H., VANHAVERBEKE W., WEST J. (eds), *Open Innovation: Researching a New Paradigm*, Oxford University Press, Oxford, 2006.

[CHI 96] CHIAPELLO È., "Les typologies des modes de contrôle et leurs facteurs de contingence : un essai d'organisation de la littérature", *Comptabilité-Contrôle-Audit*, vol. 2, no. 2, pp. 51–74, 1996.

[CHR 15] CHRONÉER D., BACKLUND F., "A holistic view on learning in project-based organizations", *Project Management Journal*, vol. 46, no. 3, pp. 61–74, 2015.

[COH 78] COHEN S., "Environmental load and the allocation of attention", in BAUM A., SINGER J.E., VALINS S. (eds), *Advances in Environmental Psychology*, vol. 1, Erlbaum, Hillsdale, 1978.

[COH 90] COHEN W.M., LEVINTHAL D.A., "Absorptive capacity: a new perspective on learning and innovation", *Administrative Science Quarterly*, vol. 35, no. 1, pp. 128–152, 1990.

[CSI 96] CSIKSZENTMIHALYI M., *Creativity: Flow and the Psychology of Discovery and Invention*, Harper Collins, New York, 1996.

[CUS 08] CUSIN J., "L'apprentissage par l'échec commercial", *Vie & sciences de l'entreprise*, vol. 1, pp. 33–53, 2008.

[CYE 63] CYERT R.M., MARCH J.G., *A Behavioral Theory of the Firm*, Prentice Hall, Engelwood Cliffs, 1963.

[DAM 14] DAMART S., Séance du 15 décembre 2014, Speech, l'Observatoire de l'innovation managériale, Paris Dauphine University, 2014.

[DAN 04] DANIEL F., LOHRKE F.T., FORNACIARI C.J. *et al.*, "Slack resources and firm performance: a meta-analysis", *Journal of Business Research*, vol. 57, no. 6, pp. 565–574, 2004.

[DEB 70] DE BONO E., *Lateral Thinking: Creativity Step by Step*, Harper & Row, 1970.

[DEM 10] DEMING P., DUNHAM R., *The Innovator's Way: Essential Practices for Successful Innovation*, MIT Press, Cambridge, 2010.

[DER 08] DEROUE D., HOLLENBECK J., JOHNSON M., "How different team downsizing approaches influence team level adaptation and performance", *Academy of Management Journal*, vol. 51, no. 1, pp. 182–196, 2008.

[DET 17] DETOURBET G., Daring to be simple: lessons learned from the Kwid, Renault-Nissan's Indian car, Speech, L'Ecole de Paris, February 13, 2017.

[DHI 08] DHIFALLAH S., CHANAL V., DEFELIX C., "Quelle gestion des ressources humaines dans les organisations ambidextres ?", *Revue française de gestion*, vol. 7, no. 187, pp. 161–175, 2008.

[DIG 90] DIGMAN J.M., "Personality structure: emergence of the five-factor model", *Annual Review of Psychology*, vol. 41, pp. 417–440, 1990.

[DOY 16] DOYEN D., FORTUNE E., La rémunération des inventions de salariés. Pratiques en vigueur en France, Analyses INPI, 2016.

[DOZ 12] DOZ Y., WILSON K., *Managing Global Innovation*, HBR Press, 2012.

[DUB 93] DUBREIL Y., "Comment réussir un projet impossible ?", *Gérer et Comprendre, Les Annales des Mines*, pp. 40–45, 1993.

[DUH 05] DUHAUTOIS R., MAUBLANC S., Les carrières des chercheurs dans les entreprises privées, Centre d'études de l'emploi, 2005.

[EDM 09] EDMONDSON A., NEMBHARD I., "Product development and learning in project teams: the challenges are the benefits", *Journal of Product Innovation Management*, vol. 26, no. 2, pp. 123–138, 2009.

[EDW 10] EDWARDS A., *Being an Expert Professional Practitioner: The Relational Turn in Expertise*, Springer, 2010.

[ERI 06] ERICSSON K.A., CHARNESS N., FELTOVICH P. *et al.*, *The Cambridge Handbook of Expertise and Expert Performance*, Cambridge University Press, 2006.

[EYH 03] EYHERABIDE M., LEFEBVRE P., ROOS P. *et al.*, *Recomposition des dynamiques professionelles des concepteurs*, Paris, 2003.

[FEI 06] FEIST G.J., *The Psychology of Science and the Origins of the Scientific Mind*, Yale University Press, New Haven, 2006.

[FRE 76] FREUD S., *Essais de psychanalyse appliquée*, Gallimard, Paris, 1976.

[FUG 05] FUGELSANG J.A., DUNBAR K.N., "Brain-based mechanisms underlying complex causal thinking", *Neuropsychologia*, vol. 43, pp. 1204–1213, 2005.

[FUG 16] FUGIER-GARREL S., MASSON R., HEINE J.-B. *et al.*, Votre R&D est-elle performante ?, Etude KL-UTC, 2016.

[GAL 83] GALAMBAUD B., *Des hommes à gérer*, Les éditions ESF, Paris, 1983.

[GAL 10] GALLIÉ E.P., FARJAUDON A.L., CHEVALIER-KUSZLA C., Les indicateurs de la R&D et l'innovation, Research report, Paris Dauphine University, 2010.

[GAL 17] GALINDO G., "À la recherche d'un idéal-type pour caractériser la GRH de la start-up high tech", *Revue de gestion des ressources humaines*, no. 103, pp. 55–70, 2017.

[GAR 03a] GAREL G., "Pour une histoire de la gestion de projet", *Gérer et Comprendre, Les Annales des Mines*, no. 74, pp. 77–89, 2003.

[GAR 03b] GAREL G., GIARD V., MIDLER C., "Management de projet et gestion des ressources humaines", in ALLOUCHE J. (ed.), *Encyclopédie de Gestion des Ressources Humaines*, Vuibert, Paris, 2003.

[GAR 03c] GAREL G., *Le management de projet*, La Découverte, Paris, 2003.

[GAS 06] GASTALDI L., GILBERT P., "Transformations du travail de recherche et GRH des chercheurs", *Revue de Gestion des Ressources Humaines*, no. 61, pp. 46–59, 2006.

[GAS 07] GASTALDI L., Stratégie d'innovation intensive et management de la recherche en entreprise, PhD thesis, Université de Marne-la-Vallée et École polytechnique, 2007.

[GAS 16] GASTALDI L., GILBERT P., "La prégnance de la double échelle de carrière experts/managers : une analyse à travers l'agence des outils de gestion", *@GRH, Numéro spécial Meilleurs papiers du 27ème congrès de l'AGRH à Strasbourg*, no. 18, pp. 11–37, 2016.

[GAU 00] GAUTIER F., GIARD V., "Vers une meilleure maîtrise des coûts engagés sur le cycle de vie, lors de la conception de produits nouveaux", *Comptabilité Contrôle Audit*, vol. 6, no. 2, pp. 43–75, 2000.

[GAU 02] GAUTIER F., Les systèmes de contrôle de gestion des projets de conception et de développement de produits nouveaux : une analyse empirique, Report, Institut d'administration des entreprises de Paris, GREGOR, 2002.

[GEE 76] GEEN R.G., "Test anxiety, observation, and range of cue utilization", *British Journal of Social and Clinical Psychology*, vol. 15, pp. 253–259, 1976.

[GEO 05] GEORGE G., "Slack resources and the performance of privately held firms", *The Academy of Management Journal*, vol. 48, no. 4, pp. 661–676, 2005.

[GIB 04] GIBSON C., BIRKINSHAW J., "The antecedents, consequences, and mediating role of organizational ambidexterity", *Academy of Management Journal*, no. 47, pp. 209–226, 2004.

[GIL 17] GILBERT P., YALENIOS J., *L'évaluation de la performance individuelle*, La Découverte, Paris, 2017.

[GIR 96] GIRAUD P.N., *L'inégalité du monde*, Gallimard, 1996.

[GIR 12] GIRAUD P.N., *La mondialisation. Emergences et fragmentations*, Editions Sciences humaines, Paris, 2012.

[GLE 16] GLEDEL F., L'apprentissage organisationnel comme outil de développement de l'innovation, Master dissertation, Aix-Marseille University, 2016.

[GOT 79] GOTZ K.O., GOTZ K., "Personality characteristics of professional artists", *Perceptual and Motor Skills*, vol. 49, pp. 327–334, 1979.

[GOU 57] GOULDNER A.W., "Cosmopolitans and locals: toward an analysis of latent social roles", *Administrative Science Quarterly*, vol. 2, no. 3, pp. 281–306, 1957.

[HAG 03] HAGEDOORN J., CLOODT M., "Measuring innovative performance: is there an advantage in using multiple indicators?", *Research Policy*, no. 32, pp. 1365–1379, 2003.

[HAR 06a] HARTMANN M., HASSAN A., "Application of real options analysis for pharmaceutical R&D project valuation – empirical results from a survey", *Research Policy*, vol. 35, no 3, pp. 343–354, 2006.

[HAR 06b] HARGADON A.B., BECHKY B.A., "When collections of creatives become creative collectives: a field study of problem solving at work", *Organization Science*, vol. 17, no. 4, pp. 484–500, 2006.

[HAT 90] HATCH M.J., "The symbolics of office design", in GAGLIARDI P. (ed.), *Symbols and Artifacts: Views of the Corporate Landscape*, Aldine de Gruyter, New York, 1990.

[HOL 12] HOLMQVIST M., SPICER A., "The ambidextrous employee: exploiting and exploring people's potential", *Research in the Sociology of Organizations*, vol. 37, pp. 1–23, 2012.

[HOO 16] HOOGE S., STASIA R., *Performance de la R&D et de l'innovation. Du contrôle de gestion à la gestion contrôlée*, Presses des Mines, Paris, 2016.

[IMM 09] IMMELT J.R, GOVINDARAJAN V., TRIMBLE C., "How GE is disrupting itself", *Harvard Business Review*, vol. 87, pp. 55–65, 2009.

[JAC 06] JACQUET D., PHILIPPE H., Les options réelles : révolution inachevée ou miroir aux alouettes ?, École de Paris du Management, Speech, December 13, 2006.

[JAC 11] JACQUIER-ROUX V., PARAPONARIS C., "L'objectif de l'internationalisation de la R&D des firmes : de la circulation au partage de connaissances tacites situées", *Management International*, vol. 16, no. 1, pp. 75–83, 2011.

[JEN 76] JENSEN M.C., MECKLING W.H., "Theory of the firm: managerial behavior, agency costs, and ownership structure", *Journal of Financial Economics*, vol. 3, pp. 305–360, 1976.

[JOL 97] JOLY P.-B., "Chercheurs et laboratoires dans la nouvelle économie de la science", *Revue d'Économie Industrielle*, no. 79, pp. 77–94, 1997.

[JOL 03] JOLIVET F., *Manager l'entreprise par projets*, EMS, Colombelles, 2003.

[KAP 96] KAPLAN R.S., NORTON D.P., *The Balanced Scorecard: Translating Strategy into Action*, Havard Business School Press, Boston, 1996.

[KAT 66] KATZ D., KAHN R.L., *The Social Psychology of Organizations*, Wiley, New York, 1966.

[KAT 04] KATZ R. (ed.), *The Human Side of Managing Technological Innovation: A Collection of Readings*, 2nd ed., Oxford University Press, Oxford, 2004.

[KRY 07] KRYCHOWSKI C., Apports et limites des options réelles à la décision d'investissement stratégique. Une étude appliquée au secteur des télécommunications, Thesis, HEC Paris, 2007.

[LAB 16] LABROUCHE G., KECHIDI M., "La stratégie d'innovation ouverte de Sanofi", *Journal de gestion et d'économie médicales*, vol. 34, pp. 65–85, 2016.

[LAP 12] LAPERCHE B., LEFEBVRE G., "The globalization of Research & Development in industrial corporations: towards reverse innovation? The cases of General Electric and Renault", *Journal of Innovation Economics & Management*, no. 10, pp. 53–79, 2012.

[LAR 10] LARSON E.W., GRAY C.F., *Project Management: The Managerial Process*, McGraw-Hill, Boston, 2010.

[LEB 12] LE BOULAIRE M., ALLOUCHE J., GILBERT P., "Le management des ressources humaines en Chine", *Encyclopédie des Ressources Humaines*, Vuibert, Paris, 2012.

[LEF 91] LEFEBVRE H., *The Production of Space* (Translated by Donaldson-Smith N.), Blackwell, Oxford, 1991.

[LEL 11] LELEBINA O., SARDAS J.-C., "L'expertise et les experts dans les organisations : une approche multidisciplinaire pour la définition des notions clés", *Congrès de l'Association Francophone de Gestion des Ressources Humaines (AGRH)*, 2011.

[LEL 14] LELEBINA O., La gestion des experts en entreprise : dynamique des collectifs de professionnels et offre de parcours, PhD thesis, Mines ParisTech, 2014.

[LEL 18] LELEBINA O., GAND S., "Beyond 'hero-based' management: revisiting HRM practices for managing collective expertise", *Relations Industrielles/Industrial Relations*, vol. 73, no. 1, pp. 39–64, 2018.

[LEM 02] LE MASSON P., WEIL B., "De la R&D à la RID : morphogénèse de la conception innovante dans les grandes entreprises : le cas securit-Saint Gobain", *Séminaire GERPISA*, 2002.

[LEM 06] LE MASSON P., WEIL B., HATCHUEL A., *Les processus d'innovation, conception innovante et croissance des entreprises*, Hermès-Lavoisier, Paris, 2006.

[LEN 02] LENFLE S., MIDLER C., "Stratégie d'innovation et organisation de la conception dans les entreprises amont", *Revue Française de Gestion*, vol. 28, no. 140, pp. 89–105, 2002.

[LEN 08] LENFLE S., "Exploration and project management", *International Journal of Project Management*, vol. 6, no. 5, pp. 469–478, 2008.

[LOR 03] LORINO P., *Méthodes et pratiques de la performance : le pilotage par les processus et les compétences*, Editions d'Organisation, Paris, 2003.

[LOU 11] LOUFRANI-FEDIDA S., "La gestion des ressources humaines au service de l'articulation entre management des compétences et organisation par projets", *Revue de gestion des ressources humaines*, vol. 1, no. 79, pp. 24–38, 2011.

[LOV 05] LOVE G., NOHRIA N., "Reducing slack: the performance consequences of downsizing by large industrial firms, 1977–93", *Strategic Management Journal*, vol. 26, pp. 1087–1108, 2005.

[MAR 03a] MARTIN D.-P., PARAPONARIS C., "Du rôle des entretiens d'appréciation dans les activités d'innovation : le cas des services R&D", *Revue de gestion des ressources humaines*, no. 50, pp. 6–22, 2003.

[MAR 03b] MARTIN D.-P., PHILLIPS N., "Organizational forgetting", in EASTERBY-SMITH M., LYLES M.A. (eds), *Handbook of Organizational Learning and Knowledge Management*, Blackwell, Oxford, 2003.

[MEL 09] MELLAHI K., WILKINSON A., "Slash and burn or nip and tuck? Downsizing, innovation and human resources", *International Journal of Human Resource Management*, vol. 21, no. 13, pp. 2291–2305, 2009.

[MEL 10] MELLAHI K., WILKINSON A., "A study of the association between level of slack reduction following downsizing and innovation output", *Journal of Management Studies*, vol. 47, no. 3, pp. 483–508, 2010.

[MEN 03] MENGER P., *Les professions et leur sociologies. Modèles théoriques, catégorisations, évolutions*, Éditions de la maison des sciences de l'homme, Paris, 2003.

[MEN 10] MENDEZ A., *Processus. Concepts et méthode pour l'analyse temporelle en sciences sociales*, Bruylant Academia, Louvain-La-Neuve, 2010.

[MID 93a] MIDLER C., "Le responsable de projet, portrait d'un rôle d'influence", *Gestion 2000*, vol. 9, pp. 123–147, 1993.

[MID 93b] MIDLER C., *L'auto qui n'existait pas*, Dunod, Paris, 1993.

[MID 17] MIDLER C., JULIEN B., LUNG Y., *Innover à l'envers*, Dunod, 2017.

[MIE 01] MIEG H.A., *The Social Psychology of Expertise: Case Studies in Research, Professional Domains and Expert Roles*, Lawrence Erlbaum Associates, Mahwah, 2001.

[MIH 10] MIHOV K., DENZLER M., FÖRSTER J., "Hemispheric specialization and creative thinking: a meta-analytic review of lateralization of creativity", *Brain and Cognition*, vol. 72, no. 3, pp. 442–448, 2010.

[MIL 15] MILOUD T., AZAN W., Evaluation de projets de R&D : Elaboration d'une méthodologie par les options réelles dans l'industrie des biotechnologies, Journée AIMS-BETA, University of Strasbourg, 2015.

[MIN 82] MINTZBERG H., *Structure et dynamique des organisations*, Éditions d'Organisation, Paris, 1982.

[MIS 98] MISHRA A.K., SPREITZER G.M., "Explaining how survivors respond to downsizing: the roles of trust, empowerment, justice, and work redesign", *The Academy of Management Review*, vol. 23, no. 3, p. 567, 1998.

[MOI 97] MOISDON J.-C., WEIL B., "Capitaliser les savoirs dans une organisation par projets", *Les Annales de l'École de Paris du management*, vol. 4, pp. 163–171, 1997.

[NAR 11] NARO G., TRAVAILLÉ D., "Le BSC à l'épreuve de l'expérience", *Revue française de gestion*, vol. 2, pp. 65–80, 2011.

[NEW 72] NEWELL A., SIMON H.A., *Human Problem Solving*, Prentice Hall, Englewood Cliffs, 1972.

[NIX 98] NIXON B., "Research and development performance measurement: a case study", *Management Accounting Research*, vol. 9, no. 3, pp. 329–355, 1998.

[NOH 96] NOHRIA N., GULATI R.A., "Is slack good or bad for organizations?", *Academy of Management Journal*, vol. 39, pp. 1245–1264, 1996.

[NON 95] NONAKA I., TAKEUCHI H., *The Knowledge Creation Company*, Oxford University Press, New York, 1995.

[OEC 02] OECD, *Frascati Manual, Proposed Standard Practise for Surveys on Research and Experimental Development*, Paris, 2002.

[ORA 16] ORANGE, Open Innovation, Internet Press file, Paris, available at: https://www.orange.com/en/content/download/10790/238197/version/2/file/Orange+Fact+Sheet+-+Open+Innovation+-+FINAL.pdf, 2016.

[PAR 00] PARAPONARIS C., "Gestion des compétences et production des connaissances dans le management par projet", *Revue de gestion des ressources humaines*, no. 36, pp. 3–17, 2000.

[PAR 18] PARAPONARIS C., CINA M., CAMPILLO V., Processus de numérisation et dynamique de l'emploi. Le cas des ingénieurs de conception industrielle, Report for APEC, LEST, Aix-en-Provence, 2018.

[PET 12] PETRONI G., VENTURINI K., VERBANO C., "Open innovation and new issues in R&D organization and personnel management", *The International Journal of Human Resource Management*, vol. 23, no. 1, pp. 147–173, 2012.

[PIC 13] PICHAULT F., NIZET J., *Les pratiques de gestion des ressources humaines*, 2nd ed., Le Seuil, 2013.

[POI 13] POINCARÉ H., *The Foundations of Science*, Science Press, Lancaster, 1913.

[RIC 15] RICHARD S., "Préface", in BEN MAHMOUD-JOUINI S., CHARUE-DUBOC F., MIDLER C. (eds), *Management de l'innovation et globalisation. Enjeux et pratiques contemporains*, Dunod, Paris, 2015.

[RIE 08] RIEU A.M., "Géostratégie de la recherche et de l'innovation", *Hermès, La Revue*, no. 50, pp. 67–73, 2008.

[ROC 94] ROCO M., "Creative personalities about creative personality in science", *Revue Roumaine de Psychologie*, vol. 37, pp. 27–36, 1994.

[ROS 03] ROSA H., *Social Acceleration: A New Theory of Modernity*, Columbia University Press, New York, 2003.

[ROS 10] ROSA H., *Accélération, une critique sociale du temps*, La Découverte, Paris, 2010.

[ROT 94] ROTHWELL R., "Toward the fifth generation innovation process", *International Marketing Review*, vol. 11, no. 1, pp. 7–31, 1994.

[ROU 98] ROUSSEAU D.M., TIJORIWALA S.A., "Assessing psychological contracts: issues, alternatives and measures", *Journal of Organizational Behavior*, vol. 19, pp. 679–695, 1998.

[RUN 99] RUNCO M.A., SAKAMOTO S.O., "Experimental research on creativity", in STERNBERG R.S. (ed.), *Handbook of Human Creativity*, Cambridge University Press, New York, 1999.

[SAC 08] SACHWALD F., "Location choices within global innovation networks: the case of Europe", *The Journal of Technology Transfer*, vol. 33, pp. 364–378, 2008.

[SCH 42] SCHUMPETER J., *Capitalism, Socialism and Democracy*, Harper & Brothers, New York, 1942.

[SHA 91] SHALLEY C.E., "Effects of productivity goals, creativity goals, and personal discretion on individual creativity", *Journal of Applied Psychology*, vol. 76, pp. 179–185, 1991.

[SIM 77] SIMON H.A., *The New Science of Management Decision*, Prentice-Hall, Englewood Cliffs, 1977.

[SIM 90] SIMONS R., "The role of management control systems in creating competitive advantage: new perspectives", *Accounting, Organizations and Society*, vol. 15, nos 1–2, pp. 127–143, 1990.

[SIM 08] SIMONI G., "Comment capitaliser les connaissances générées par les projets de R&D ?", *Gérer et comprendre, Les Annales des Mines*, vol. 91, no. 1, pp. 67–78, 2008.

[STÄ 10] STÄRK M., Strategies, management and HR practices within companies in China, Speech, CIRAC, Paris, June 4, 2010.

[SUS 07] SUSSKIND A.M., "Downsizing survivors' communication networks and reactions: a longitudinal examination of information flow and turnover intentions", *Communication Research*, vol. 34, pp. 144–156, 2007.

[TAK 86] TAKEUCHI H., NONAKA I., "The new product development game", *Harvard Business Review*, vol. 64, pp. 137–146, 1986.

[TID 05] TIDD J., BESSANT J., PAVITT K., *Managing Innovation Integrating Technological, Market and Organizational Change*, Wiley, 2005.

[TRI 13] TRIOMPHE C., "Management de projets : rationalisation ou agilité ?", in SAUSSIER S. (ed.), *Management des entreprises*, Eyrolles, Paris, 2013.

[TUS 96] TUSHMAN M., O'REILLY C., "Ambidextrous organizations: managing evolutionary and revolutionary change", *California Management Review*, no. 38, pp. 8–30, 1996.

[VON 01] VON HIPPEL E., "Perspective: user toolkits for innovation", *Journal of Product Innovation Management*, vol. 18, no. 4, pp. 247–257, 2001.

[VON 13] VON HIPPEL E., "User innovation", in HUFF A.S., MOSLEIN K.M., REICHWALD R. (eds), *Leading Open Innovation*, MIT Press, Cambridge, 2013.

[WAL 26] WALLAS G., *The Art of Thought*, Harcourt Brace, New York, 1926.

[WEG 86] WEGNER D.M., "Transactive memory: a contemporary analysis of the group mind", in MULLEN B., GOETHALS G. (eds), *Theories of Group Behavior*, Springer-Verlag, New York, 1986.

[WEI 08] WEINMANN N., R&D des compagnies pharmaceutiques : ruptures et mutations, Mission Prospective pour le compte de la Direction Générale des Entreprises, Observatoire des Stratégies Industrielles, Paris, 2008.

[WIL 81] WILLIAMSON O.E., "The economics of organization: the transaction cost approach", *The American Journal of Sociology*, vol. 87, no. 3, pp. 548–577, 1981.

[YLI 09] YLI-KAUHALUOMA S., "Time at R&D Work: types and strategies of time in the collaborative development of a chemical technology", *Time & Society*, vol. 18, no. 1, pp. 130–153, 2009.

[ZAN 08] ZANNAD H., "La gestion des ressources humaines dans les projets industriels", *Revue de gestion des ressources humaines*, vol. 2, no. 68, pp. 49–65, 2008.

Index

O, P, R

S, T, U, W

Other titles from

ISTE

in

Innovation, Entrepreneurship and Management

2018

CALLENS Stéphane
Creative Globalization
(Smart Innovation Set – Volume 16)

CASADELLA Vanessa
Innovation Systems in Emerging Economies: MINT – Mexico, Indonesia, Nigeria, Turkey
(Smart Innovation Set – Volume 18)

CORLOSQUET-HABART Marine, JANSSEN Jacques
Big Data for Insurance Companies
(Big Data, Artificial Intelligence and Data Analysis Set – Volume 1)

CROS Françoise
Innovation and Society
(Smart Innovation Set – Volume 15)

DEBREF Romain
Environmental Innovation and Ecodesign: Certainties and Controversies
(Smart Innovation Set – Volume 17)

DOMINGUEZ Noémie
SME Internationalization Strategies: Innovation to Conquer New Markets

ERMINE Jean-Louis
Knowledge Management: The Creative Loop
(Innovation and Technology Set – Volume 5)

IBRAHIMI Mohammed
Mergers & Acquisitions : Theory, Strategy, Finance

LÉVY Aldo, BEN BOUHENI Faten, AMMI Chantal
Financial Management: USGAAP and IFRS Standards
(Innovation and Technology Set – Volume 6)

SACHETTI Philippe, ZUPPINGER Thibaud
New Technologies and Branding
(Innovation and Technology Set – Volume 4)

TEMPLE Ludovic, COMPAORÉ SAWADOGO Eveline M.F.W.
Innovation Processes in Agro-Ecological Transitions in Developing Countries
(Innovation in Engineering and Technology Set – Volume 2)

VAN HOOREBEKE Delphine
The Management of Living Beings or Emo-management

2017

AÏT-EL-HADJ Smaïl
The Ongoing Technological System
(Smart Innovation Set – Volume 11)

BAUDRY Marc, DUMONT Béatrice
Patents: Prompting or Restricting Innovation?
(Smart Innovation Set – Volume 12)

BÉRARD Céline, TEYSSIER Christine
Risk Management: Lever for SME Development and Stakeholder Value Creation

CHALENÇON Ludivine
Location Strategies and Value Creation of International Mergers and Acquisitions

CHAUVEL Danièle, BORZILLO Stefano
The Innovative Company: An Ill-defined Object
(Innovation Between Risk and Reward Set – Volume 1)

CORSI Patrick
Going Past Limits To Growth

D'ANDRIA Aude, GABARRET Inés
Building 21st Century Entrepreneurship
(Innovation and Technology Set – Volume 2)

DAIDJ Nabyla
Cooperation, Coopetition and Innovation
(Innovation and Technology Set – Volume 3)

FERNEZ-WALCH Sandrine
The Multiple Facets of Innovation Project Management
(Innovation between Risk and Reward Set – Volume 4)

FOREST Joëlle
Creative Rationality and Innovation
(Smart Innovation Set – Volume 14)

GUILHON Bernard
Innovation and Production Ecosystems
(Innovation between Risk and Reward Set – Volume 2)

HAMMOUDI Abdelhakim, DAIDJ Nabyla
Game Theory Approach to Managerial Strategies and Value Creation
(Diverse and Global Perspectives on Value Creation Set – Volume 3)

LALLEMENT Rémi
Intellectual Property and Innovation Protection: New Practices and New Policy Issues
(Innovation between Risk and Reward Set – Volume 3)

LAPERCHE Blandine
Enterprise Knowledge Capital
(Smart Innovation Set – Volume 13)

LEBERT Didier, EL YOUNSI Hafida
International Specialization Dynamics
(Smart Innovation Set – Volume 9)

MAESSCHALCK Marc
Reflexive Governance for Research and Innovative Knowledge
(Responsible Research and Innovation Set – Volume 6)

MASSOTTE Pierre
Ethics in Social Networking and Business 1: Theory, Practice and Current Recommendations
Ethics in Social Networking and Business 2: The Future and Changing Paradigms

MASSOTTE Pierre, CORSI Patrick
Smart Decisions in Complex Systems

MEDINA Mercedes, HERRERO Mónica, URGELLÉS Alicia
Current and Emerging Issues in the Audiovisual Industry
(Diverse and Global Perspectives on Value Creation Set – Volume 1)

MICHAUD Thomas
Innovation, Between Science and Science Fiction
(Smart Innovation Set – Volume 10)

PELLÉ Sophie
Business, Innovation and Responsibility
(Responsible Research and Innovation Set – Volume 7)

SAVIGNAC Emmanuelle
The Gamification of Work: The Use of Games in the Workplace

SUGAHARA Satoshi, DAIDJ Nabyla, USHIO Sumitaka
Value Creation in Management Accounting and Strategic Management: An Integrated Approach
(Diverse and Global Perspectives on Value Creation Set –Volume 2)

UZUNIDIS Dimitri, SAULAIS Pierre
Innovation Engines: Entrepreneurs and Enterprises in a Turbulent World
(Innovation in Engineering and Technology Set – Volume 1)

2016

BARBAROUX Pierre, ATTOUR Amel, SCHENK Eric
Knowledge Management and Innovation
(Smart Innovation Set – Volume 6)

BEN BOUHENI Faten, AMMI Chantal, LEVY Aldo
Banking Governance, Performance And Risk-Taking: Conventional Banks Vs Islamic Banks

BOUTILLIER Sophie, CARRÉ Denis, LEVRATTO Nadine
Entrepreneurial Ecosystems (Smart Innovation Set – Volume 2)

BOUTILLIER Sophie, UZUNIDIS Dimitri
The Entrepreneur (Smart Innovation Set – Volume 8)

BOUVARD Patricia, SUZANNE Hervé
Collective Intelligence Development in Business

GALLAUD Delphine, LAPERCHE Blandine
Circular Economy, Industrial Ecology and Short Supply Chains
(Smart Innovation Set – Volume 4)

GUERRIER Claudine
Security and Privacy in the Digital Era
(Innovation and Technology Set – Volume 1)

MEGHOUAR Hicham
Corporate Takeover Targets

MONINO Jean-Louis, SEDKAOUI Soraya
Big Data, Open Data and Data Development
(Smart Innovation Set – Volume 3)

MOREL Laure, LE ROUX Serge
Fab Labs: Innovative User
(Smart Innovation Set – Volume 5)

PICARD Fabienne, TANGUY Corinne
Innovations and Techno-ecological Transition
(Smart Innovation Set – Volume 7)

2015

CASADELLA Vanessa, LIU Zeting, DIMITRI Uzunidis
Innovation Capabilities and Economic Development in Open Economies
(Smart Innovation Set – Volume 1)

CORSI Patrick, MORIN Dominique
Sequencing Apple's DNA

CORSI Patrick, NEAU Erwan
Innovation Capability Maturity Model

FAIVRE-TAVIGNOT Bénédicte
Social Business and Base of the Pyramid

GODÉ Cécile
Team Coordination in Extreme Environments

MAILLARD Pierre
Competitive Quality and Innovation

MASSOTTE Pierre, CORSI Patrick
Operationalizing Sustainability

MASSOTTE Pierre, CORSI Patrick
Sustainability Calling

2014

Dubé Jean, Legros Diègo
Spatial Econometrics Using Microdata

Lesca Humbert, Lesca Nicolas
Strategic Decisions and Weak Signals

2013

Habart-Corlosquet Marine, Janssen Jacques, Manca Raimondo
VaR Methodology for Non-Gaussian Finance

2012

Dal Pont Jean-Pierre
Process Engineering and Industrial Management

Maillard Pierre
Competitive Quality Strategies

Pomerol Jean-Charles
Decision-Making and Action

Szylar Christian
UCITS Handbook

2011

Lesca Nicolas
Environmental Scanning and Sustainable Development

Lesca Nicolas, Lesca Humbert
Weak Signals for Strategic Intelligence: Anticipation Tool for Managers

Mercier-Laurent Eunika
Innovation Ecosystems

2010

SZYLAR Christian
Risk Management under UCITS III/IV

2009

COHEN Corine
Business Intelligence

ZANINETTI Jean-Marc
Sustainable Development in the USA

2008

CORSI Patrick, DULIEU Mike
The Marketing of Technology Intensive Products and Services

DZEVER Sam, JAUSSAUD Jacques, ANDREOSSO Bernadette
Evolving Corporate Structures and Cultures in Asia: Impact of Globalization

2007

AMMI Chantal
Global Consumer Behavior

2006

BOUGHZALA Imed, ERMINE Jean-Louis
Trends in Enterprise Knowledge Management

CORSI Patrick *et al.*
Innovation Engineering: the Power of Intangible Networks

Printed and bound by CPI Group (UK) Ltd, Croydon, CR0 4YY
15/09/2026
14962349-0003